i

Stress Free™ Daily Management Solutions
Zero Defect, Twenty-Four Hour, Management

By: Ron Mueller

Around the World Publishing LLC
4914 Cooper Road Suite 144
Cincinnati, Ohio 45242-9998

ISBN 13: 978-1-68223-124-1

Distributed by Ingram
Cover Picture By: Drazen Zigic, Shutterstock.com
Cover Design By: Ron Mueller

Ron Mueller

CONTENTS

DEDICATION

To all the people that want to make it
a stable, organized execution of daily work
and to make work life manageable

Ron Mueller

ACKNOWLEDGMENTS

To all the people that want a stable and well-organized workday and want to make their work life more manageable.

TECHNICAL EDITOR:
Gordon Miller P. E.

Introduction

Daily Management

Daily management is by definition the management of the work done by people on multiple teams and by multiple organizations engaged at several organizational levels in the transformation of materials and information into products desired by the customer and the production of organization reports desired by the organization leaders.

Maintain and Improve

There are two sides to Daily Management.

There is Daily Work Management (D.W.M.) and there is the Daily Continuous Improvement (D.C.I.).

Daily Work Management (D.W.M.) focuses on Maintaining the work environment at a zero loss, zero waste condition. It is always focused on ensuring that the next twenty-four hours are at standard.

Daily Continuous Improvement (D.C.I.) focuses on the creation of new skills, improvement of the supply chain transformation elements and improving the customer's satisfaction by providing what is desired when it is desired and where it is desired.

What is Stress Free™ Daily Management Solutions?

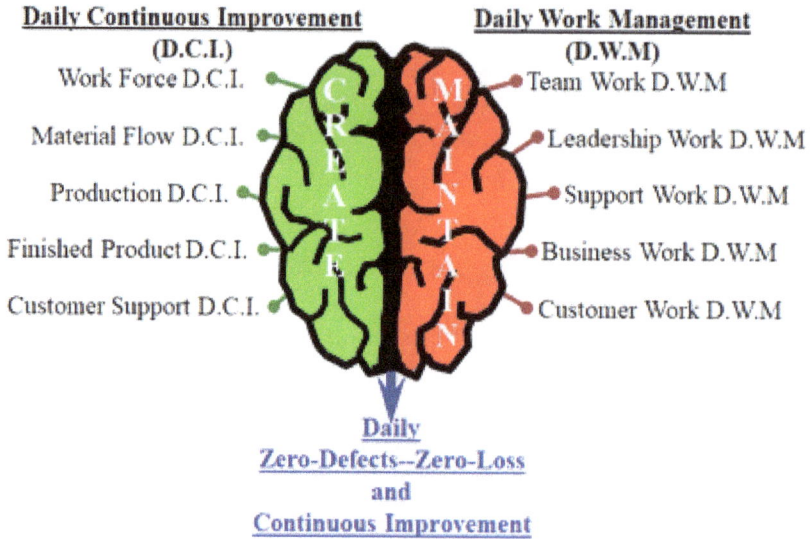

Daily Continuous Improvement (D.C.I.)	Daily Work Management (D.W.M)
Work Force D.C.I.	Team Work D.W.M
Material Flow D.C.I.	Leadership Work D.W.M
Production D.C.I.	Support Work D.W.M
Finished Product D.C.I.	Business Work D.W.M
Customer Support D.C.I.	Customer Work D.W.M

CREATE **MAINTAIN**

**Daily
Zero-Defects--Zero-Loss
and
Continuous Improvement**

Daily Management Goal

The goal of zero product defects, zero material waste and zero lost time for the next twenty-four hours is achieved by utilizing an organized daily management method detailed in Stress Free™ Daily Management Solutions.

This daily management is based on the participation of the entire organization. It requires leadership coordination and adherence to standards.

Leadership begins on the line operating team. Each operating team member has a secondary role along with their line operation role. Operating team members assume a secondary role such as, Safety coordinator, Quality coordinator, Throughput tracker, Maintenance coordinator, OEE tracker, Team leader.

The Operating team leaders, one per shift, utilize the output production tracking sheet to understand the operational situation and communicate with the next shift and the department manager.

The use of a standard agenda and the use of the Production Output Tracking sheet supports effective and efficient meetings up through the organization's leadership.

Daily Management Value

The stability generated by an organization practicing daily management generates zero losses for each 24-hour operating period, resulting in millions of dollars savings all along the entire supply chain.

Value Elements

The primary purpose of business is to make money. The specific elements of time, materials, production, finished product and consumer satisfaction are all involved in creating the business value or generating the desired profit. A stable production system supports these business goals.

Time

Time is an element in all the actions and transformations that occur along the supply chain. It is the metric that allows the measurement of consistency. Think of time as the universal element: the time to bring in raw materials, the time to handle the raw materials, the time to transport the raw materials to the point of transformation, the time to transform the raw materials into a finished product, the time to move the finished product to a temporary holding area, the time to retrieve and load the finished product onto the transport vehicle, the time to transport the finished product to the buying customer, the time to put the finished product into the hands of the final consumer.

Time is the fundamental unit.

Materials

The materials to make a product need to come together in an orchestrated way that meets the flow of each material moving into the production realm. It is important to synchronize each material with the flow required by the transformation process. The application of the Incoming Material Tracking tool highlights issues with key materials needed for production.

Production

The transformation of materials at the customer product pull rate results in the ability to hold a cost-effective inventory level. It also supports stable and consistent production work practices.

The primary tools are; Production Output Tracking, Standard Daily Management meeting agendas, use of visual signals and a clear split of production and maintenance responsibilities.

Finished Product Handling

Holding a three-sigma level (this level assures that 99.7% of the time there will be enough inventory) of finished product inventory and handling the inventory a minimal number of times contributes to finished product stability. This stability reflects the condition of all the work processes and actions up stream in the supply chain.

The ability to deliver the finished product to the customer, on time, in the quantity desired, at the point desired completes the business cycle.

Finished Product Output tracking aids in creating a stable controllable system.

Customer Satisfaction

A satisfied customer is the goal of every business. The Stress Free™ approach asks the customer what they want and then it meets the customer requirement and generates customer satisfaction and a desire for continued partnership.

Output and Input tracking utilized at partner boundaries provides a simple, common language way to understand the flow issues.

Output tracking is explained in detail in chapter eight.

Achieved Value

The Stress Free™ approach, tools and techniques consistently deliver the desired results. Time, material and finished product inventory is reduced. Production throughput and finished product handling is synchronized to the customer pull. The customer in all cases is more satisfied and develops a closer working relationship with you, their supplier.

Daily Management Tools (See chapter 8)

The tools used to create and maintain Stress Free™ Daily Management Solutions are a set of simple techniques, meetings, team and leadership practices. These elements used in a specific way and managed in a specific time sequence result in a supply chain flow that experiences minimum loss and achieves high stability. The people implementing the techniques and tools all experience work process stability and error free work that results in an environment that invigorates them.

Daily Management Implementation

Some key considerations when implementing Stress Free™ Daily Management Solutions are; the organization design, team-oriented work processes, just in time and mistake proof practices, no over production and meeting the customer expectations.

The Production Area Personnel at the top.

Daily Management begins with the production line operating teams. The production operating teams ensure that the material transformation leading to the product desired by the purchasing customer is defect free at the

What is Stress Free™ Daily Management Solutions?

minimum loss possible for the current system. These teams take part in Daily Work Management (D.W.M) and in Daily Continuous Improvement (D.C.I.)

Daily Management Organization Structure Example

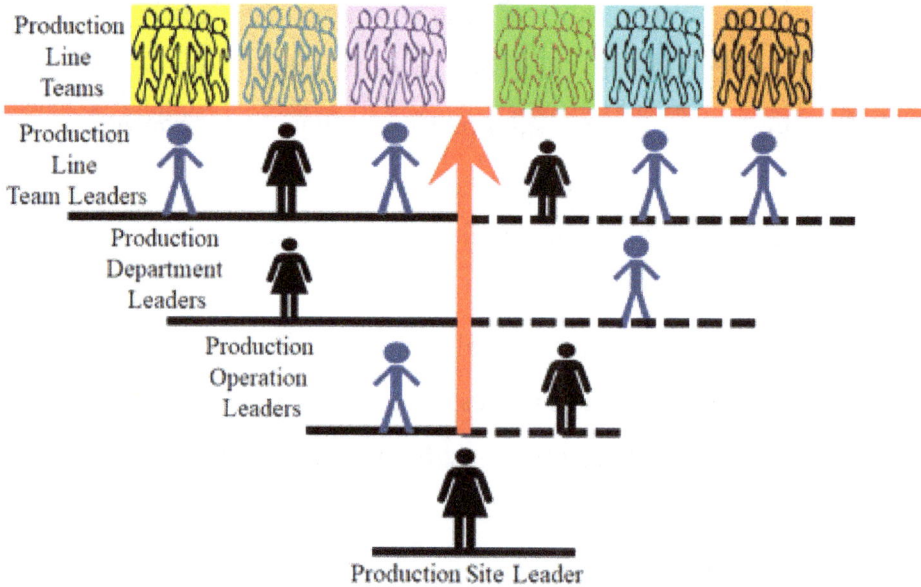

Production Line Teams

Production Line Team Leaders

Production Department Leaders

Production Operation Leaders

Production Site Leader

Each production line team is supported by the production line team leader. This team leader is an integral member of the production line team but has earned the role of leading the team. He or she makes sure the team works to standard and the team continues to develop their skills.

Production line teams are in turn supported by the Production Department Leader. The Production Department Leader normally represents the first organizational level that manages a spend budget. This budget is largely focused on maintaining the production transformation processes. This leader is often the first level responsible for the entire line for every twenty-four-hour period.

Production Department Leaders are supported by Production Operations Leaders. The Production Operations Leader manages personnel assignment and often directly ensures the Daily Continuous Improvement (DCI) for the work force, the material flow and finished product flow.

The role of supporting Finished Product Handling is normally a logistics role at the Operation level and includes handling the storage and the shipment of the finished product.

The Customer Support is most often handled by the production site leader in conjunction with a customer sales representative.

Support Organizations, Safety, Quality, Engineering, Accounting participate in the Daily Production Management activities in a support role and then hold their own Daily Management meeting.

The size of the organization determines the number of organizational levels and the responsibilities of each level. The five levels described will fit all medium to large production facilities. Smaller organization normally still required a solid daily management system. The main difference is that the people in the smaller organization wear multiple responsibility hats.

The application of Stress Free™ Daily Management Solutions applies to all organization sizes.

Chapter 1: Shift Team Daily Management

The shift operating team operates and maintains the production line. They are responsible for a flawless shift. It is important for the operating team to have a successful, flawless shift and to ensure that the next shift is set up to continue the flawless operation and maintenance of the production line.

Daily Management Activities

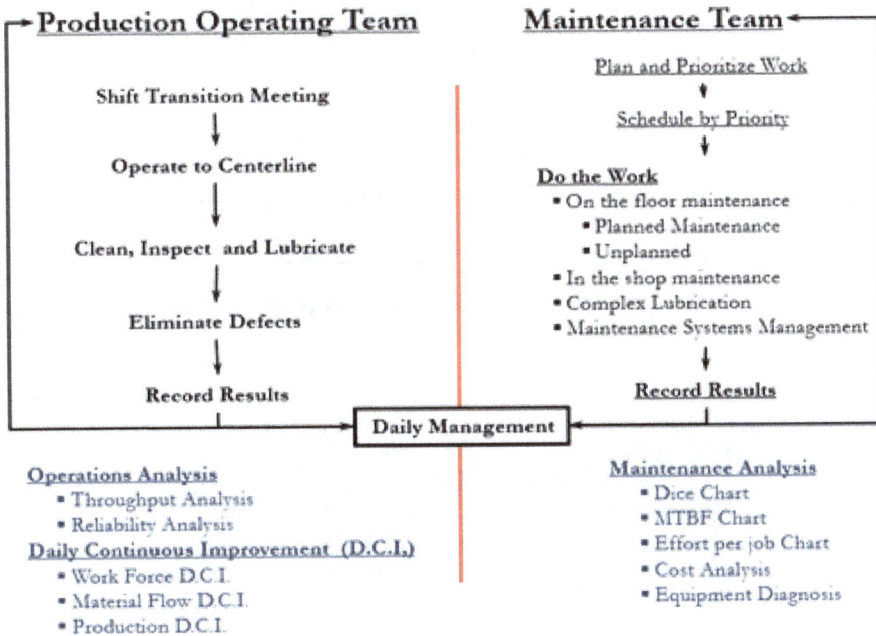

Production Operating Team	Maintenance Team
	Plan and Prioritize Work
	↓
	Schedule by Priority
Shift Transition Meeting	↓
↓	Do the Work
Operate to Centerline	• On the floor maintenance
↓	• Planned Maintenance
	• Unplanned
Clean, Inspect and Lubricate	• In the shop maintenance
↓	• Complex Lubrication
Eliminate Defects	• Maintenance Systems Management
↓	↓
Record Results	Record Results

Daily Management

Operations Analysis
- Throughput Analysis
- Reliability Analysis

Daily Continuous Improvement (D.C.I.)
- Work Force D.C.I.
- Material Flow D.C.I.
- Production D.C.I.

Maintenance Analysis
- Dice Chart
- MTBF Chart
- Effort per job Chart
- Cost Analysis
- Equipment Diagnosis

Production Line Operation Team

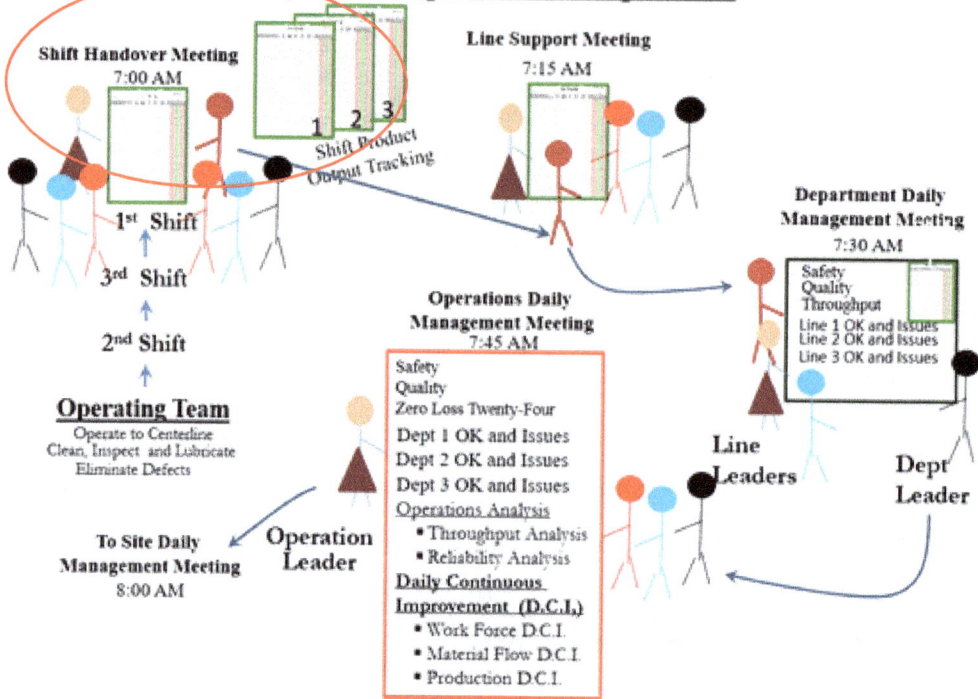

Shift Hand Over Meeting

The shift handover meeting is the beginning of a sequence that flows up to the production operation level. Each organizational level utilizes a similar agenda. This supports a common language and similar culture throughout the organization.

The shift hand over meeting is focused on the off going shift updating the oncoming shift. The meeting is intended to be focused and short. Output tracking puts all the information on one sheet. Technology is changing how the output tracking sheet is shared. It is now feasible to do this via the i-phone.

This meeting format is followed up to the Operation level. Each organizational level will add items to the agenda for which they are responsible.

The oncoming team leader reviews the previous two shifts output tracking. This provides context for what happened during the time the shift team had been off.

The outgoing production operating team leader conducts a transition meeting with the oncoming production operating team. The exchange focuses on any safety and quality issues as well as the production system reliability and throughput achieved. If improvements have been made this is also communicated.

The shift production output tracking chart is the focus of this discussion. The output tracking chart shows the hour by hour scheduled and achieved production volume. Hourly notes capture the production system performance. The line reliability is tracked on an hour by hour basis.

It is critical that each of the meetings is focused and in the fifteen-minute range.

Output Tracking

The output tracking sheet is a comprehensive hour by hour presentation of the line production.

Output Tracking										

Area: A+B+C+Applicator
Team: Customization team
Shift: Day shift
Area Leader: Maria Huyhn

Total Target Flow	980
Produced to Schedule	980
Produced Overall	980

Available Time	7 hours
Cycle Time	30 sec/person
Unit/ Min	2 boxes

Simple OEE	100.0%

Center-Line

Time hr.	Product	Target Flow	Actual Flow	Difference	Leader Initials	Issue and Root Cause	Root Cause Responsibility	by Date	Stability Tracking −	+
1	Facial Treatment Package	140	140	0						
2	Facial Treatment Package	140	135	(5)		Package scratch. Quality Issue				
3	Facial Treatment Package	140	140	0						
4	Facial Treatment Package	0	0	0						
5	Facial Treatment Package	140	140	0						
6	Facial Treatment Package	140	140	0						
7	Facial Treatment Package	140	140	0						
8	Facial Treatment Package	140	145	5						
9										
10										
11										
12										

Shift Summary
Five quality defect. Made up at last hour 8.
Shift ended with complete scheduled production

Output Tracking forms the basis of communication along the supply chain and up through the organization. The daily meeting agenda and format utilizes output tracking as a basis for a common discussion.

Output or input tracking should be utilized at every supply chain boundary. Output tracking is a simple form that provides the participants on two sides of an organizational or supply chain boundary to communicate clearly. It provides the information that results in reducing the problems at every boundary.

Products of Chapter 1:

1. Aligned focused operating team
2. Safe operation
3. Stability and product produced at Quality
4. Zero Loss next twenty-four hours.

Chapter 1 Tips and Tools

1. Use Standard Meeting Agenda
2. Use output tracking
3. Use Next Twenty-Four Task List
4. Schedule Support help

Chapter 2: Line Team Leader Support Meeting

The line leader meets with the line support members to review the issues facing the operating team and to determine if the line support personnel agree that they are able to handle the identified issues. The intent is to make sure that the line has the required support for the next upcoming shift.

Daily Management Sequence

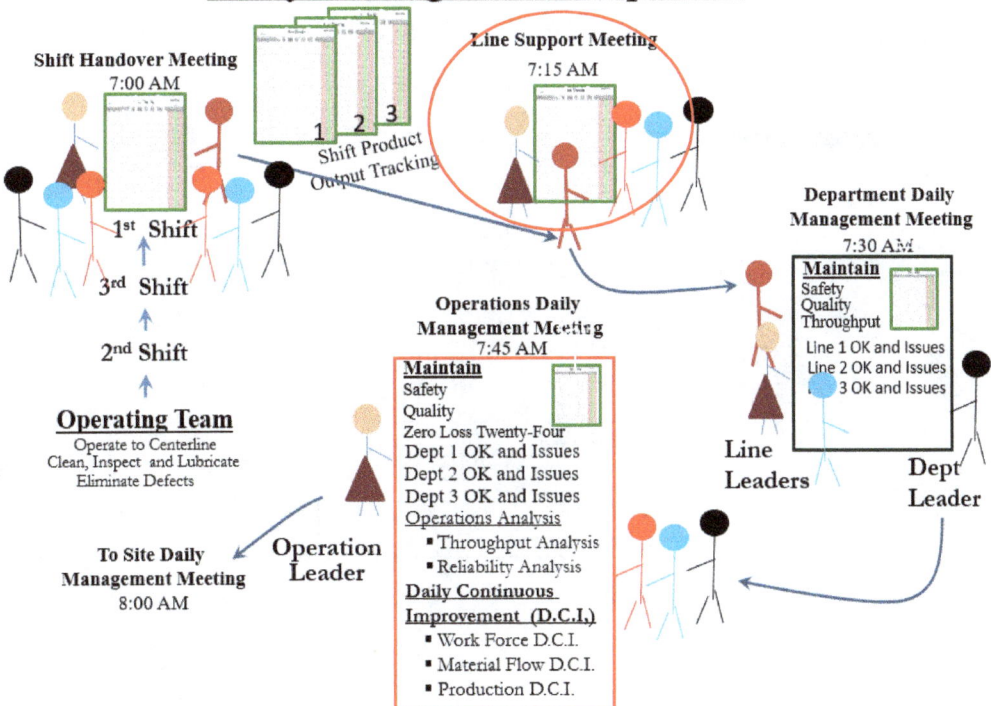

The agenda that is followed is the same as the line shift team used. This makes for a quick transition for the line leader in the discussions with the support team.

Line Support Meeting		
OBJECTIVE: - Review last 24 hrs results to identify top 3 losses, determine immediate actions and root cause actions. - Determine and additional actions for shift team. - Verify that the support resources are able to address any outstanding issues		
Focus	**Who**	**Information Needed**
Incident	Line Leader	Incident forms
Unsafe condition & behaviour	Line Leader	Verbal information
Behavior Observation Completion	Line Leader	Verbal information
Quality Issues	Line Leader	Incident forms & sample
Parts Per Million Defects	Line Leader	Output Tracking
Total Scrap	Line Leader	Output Tracking
Unplanned stop (#)	Line Leader	Output Tracking
Unplanned PR Loss (%)	Line Leader	Output Tracking
Previous Shift Production	Line Leader	Output Tracking
Production Planned for the upcoming shift	Line Leader	Output Tracking
Changer Over (C/O) plan and time	Line Leader	Output Tracking
Clean Inspect and Lubricate (CIL) plan	Line Leader	CIL history
Current Team (CIL) Completion	Line Leader	CIL history
False Starts	Line Leader	False Starts Report
Maintenance Requirements	Line Leader	Maintenance Schedule

PARTICIPANTS
1. Line Leader   4. Process Technician
2. Maintenance Technician   5. Department Safety Owner (as needed)
3. E&I Technician   6. Department Quality Owner (as needed)

The Line support meeting allows the line leader to marshal the resources needed to address any of the current issues facing the line.

Critical Item Task List

Critical Item	Safety	Quality	Throughput	Criticality Value
Flashes at the neck- Quality	4	3	8	96
Rounded internal edge around the neck- Quality	3	8	5	120
Less than 2 mm internal sealing zone at the neck- Quality	4	8	6	288
Bulging of the bottle-Quality	9	6	3	162
Flashes around the bottom-Quality	4	6	7	168
Orange peel effect-Quality	3	1	1	3
Holes in the bottle-Quality	3	5	2	30
Denting on the bottom part-Quality	8	8	4	256
Impurities	10	8	7	560
zz				1

Critical Items Priority Parato

Values shown on chart: 560, 288, 256, 168, 162, 120, 96, 30, 3, 1

Rating	Priority	Item	Priority	Resolution Owner
1=	Low Priority	Impurities	560	
9=	Low Priority	Less than 2 mm internal sealing zone at the neck-Quality	288	
27=	Low Priority	Denting on the bottom part-Quality	256	
64=	Low Priority	Flashes around the bottom-Quality	168	
125=	Medium	Bulging of the bottle-Quality	162	
343=	Medium	Rounded internal edge around the neck- Quality	120	
512=	High	Flashes at the neck- Quality	96	
729=	Very High	Holes in the bottle-Quality	30	
1000=	Critical	Orange peel effect-Quality	3	
		zz	1	

A key product is the Critical Item Task list. The critical items are listed, and resolution owners assigned. This focus is aimed at producing a zero loss for the next twenty-four-hours. The items are rated as to criticality and then action is taken based on priority.

The coordination and sharing of information with the support personnel eliminates surprise situations. It also determines if the any item on the critical list requires resource beyond the line support team.

Maintenance Support

The maintenance organization is a critical partner with the line operating team. They provide equipment training, equipment maintenance and critical lubrication and coach the operating team how to do the less difficult lubrication when the line is down for cleaning.

The maintenance organization supports the operating team when significant maintenance is required. The degree of maintenance support is determined by the experience of the shift operating team and the condition of the production equipment. An example maintenance plan is shown.

		Maintence Plan 2020												Maintence Plan 2021											
Operation 1	Equipment	January	February	March	April	May	June	July	August	September	October	November	December	January	February	March	April	May	June	July	August	September	October	November	December
Department 1																									
	Line 1	H	Major OH	s	s	s	s	s	s	s	s	s	s	H	s	s	Major OH	s	s	s	s	s	s	s	s
	Line 2	s	H	Major OH	s	s	s	s	s	s	s	s	s	s	H	s	Major OH	s	s	s	s	s	s	s	s
	Line 3	s	s	H	Major OH	s	s	s	s	s	s	s	s	s	s	H	Major OH	s	s	s	s	s	s	s	s
Department 2																									
	Line 1	s	s	s	s	H	Major OH	s	s	s	s	s	s	s	s	s	s	H	s	s	Major OH	s	s	s	s
	Line 2	s	s	s	s	s	H	Major OH	s	s	s	s	s	s	s	s	s	s	H	s	Major OH	s	s	s	s
	Line 3	s	s	s	s	s	s	H	Major OH	s	s	s	s	s	s	s	s	s	s	H	Major OH	s	s	s	s
Department 3																									
	Line 1	s	s	s	s	s	s	s	H	Major OH	s	s	s	s	s	s	s	s	s	s	s	H	s	s	Major OH
	Line 2	s	s	s	s	s	s	s	s	H	Major OH	s	s	s	s	s	s	s	s	s	s	s	H	s	Major OH
	Line 3	s	s	s	s	s	s	s	s	s	s	H	Major OH	s	s	s	s	s	s	s	s	s	s	H	Major OH
Operation 2	Equipment	January	February	March	April	May	June	July	August	September	October	November	December	January	February	March	April	May	June	July	August	September	October	November	December
Department 1																									
	Line 1	Major OH	H	s	s	s	s	s	s	s	s	s	s	Major OH	H	s	s	s	s	s	s	s	s	s	s
	Line 2	s	Major OH	H	s	s	s	s	s	s	s	s	s	s	Major OH	H	s	s	s	s	s	s	s	s	s
	Line 3	s	s	Major OH	H	s	s	s	s	s	s	s	s	s	s	Major OH	H	s	s	s	s	s	s	s	s
Department 2																									
	Line 1	s	s	s	s	Major OH	H	s	s	s	s	s	s	s	s	s	s	Major OH	H	s	s	s	s	s	s
	Line 2	s	s	s	s	s	Major OH	H	s	s	s	s	s	s	s	s	s	s	Major OH	H	s	s	s	s	s
	Line 3	s	s	s	s	s	s	Major OH	H	s	s	s	s	s	s	s	s	s	s	Major OH	H	s	s	s	s
Department 3																									
	Line 1	s	s	s	s	s	s	s	Major OH	H	s	s	s	s	s	s	s	s	s	s	Major OH	H	s	s	s
	Line 2	s	s	s	s	s	s	s	s	Major OH	H	s	s	s	s	s	s	s	s	s	s	Major OH	H	s	s
	Line 3	s	s	s	s	s	s	s	s	s	Major OH	H	s	s	s	s	s	s	s	s	s	s	Major OH	H	s

Major OH = Production Line Down for a minimum of one week
H = Production Line Down for 24 Hrs
s = Support as needed

The goal is to maintain the line at its peak operating condition. For this to occur maintenance work needs to be closely coordinated with the operation. It is important to schedule major maintenance work, just as it is important to schedule product production.

The maintenance goal is to manage maintenance time in such a way as to optimize the production time.

17

The production line operator skills are improved to the point that they do the normal line lubrication and minor repairs. This is a key part of growing the production line skills. This operator hands on approach to line maintenance dramatically improves production line performance.

Output tracking captures this performance on an hour by hour basis.

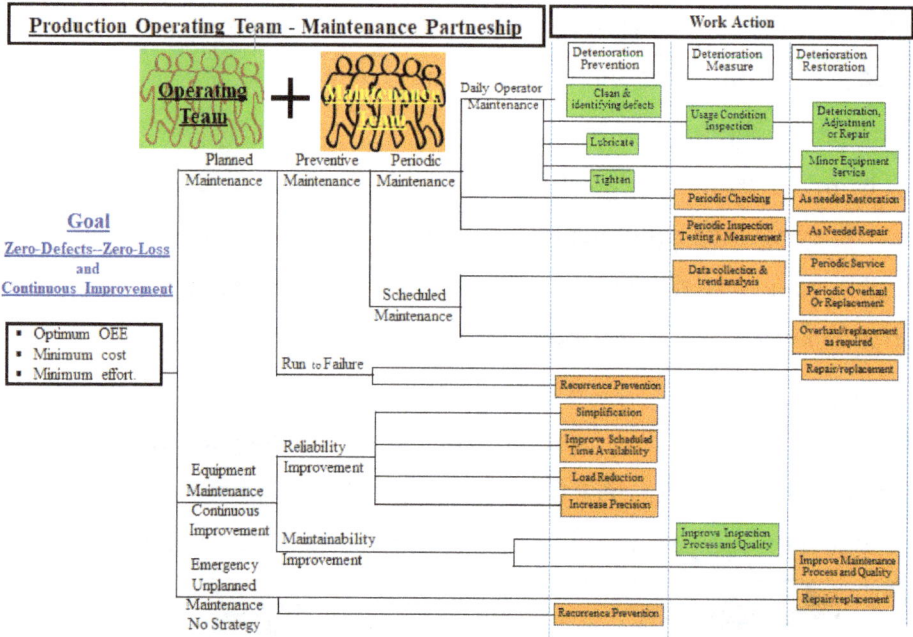

Chapter 3: Department Daily Management

Daily Management Sequence

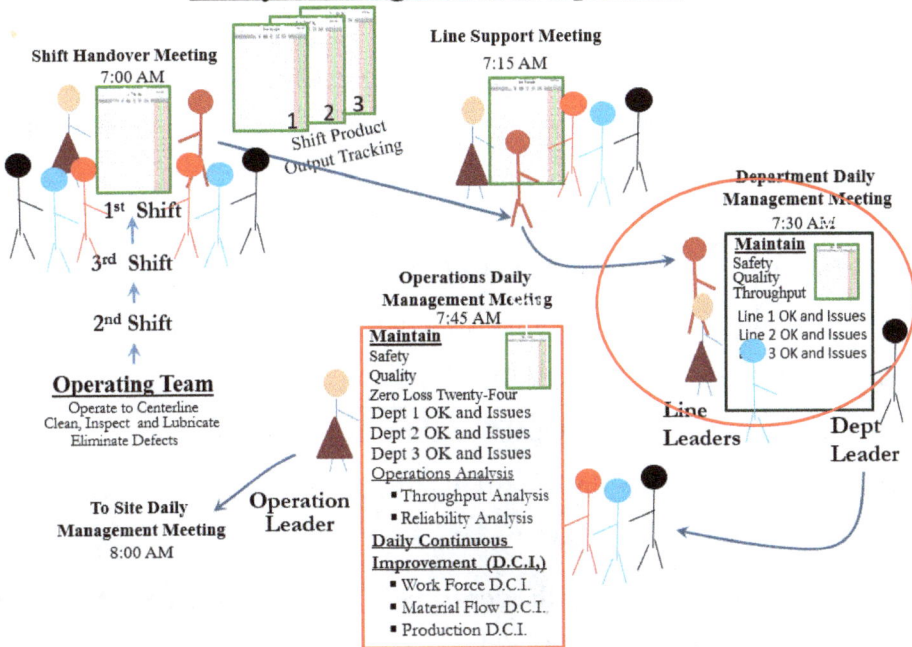

Shift Handover Meeting
7:00 AM

Line Support Meeting
7:15 AM

1 2 3
Shift Product
Output Tracking

1st Shift

3rd Shift

2nd Shift

Operating Team
Operate to Centerline
Clean, Inspect and Lubricate
Eliminate Defects

Department Daily Management Meeting
7:30 AM

Maintain
Safety
Quality
Throughput
Line 1 OK and Issues
Line 2 OK and Issues
Line 3 OK and Issues

Line
Leaders

Dept
Leader

Operations Daily Management Meeting
7:45 AM

Maintain
Safety
Quality
Zero Loss Twenty-Four
Dept 1 OK and Issues
Dept 2 OK and Issues
Dept 3 OK and Issues
Operations Analysis
▪ Throughput Analysis
▪ Reliability Analysis
Daily Continuous Improvement (D.C.I.)
▪ Work Force D.C.I.
▪ Material Flow D.C.I.
▪ Production D.C.I.

Operation Leader

To Site Daily Management Meeting
8:00 AM

The department daily meeting comes after the operating team leaders support meetings. All the Line leaders are done reviewing the previous shift performance utilizing the Production Output Tracking Charts. They understand the condition of their line. They are clear about their needs. They have evaluated the ability of the line to meet the upcoming shift's needs.

Department Daily Meeting Agenda

OBJECTIVE: - To align on departmental priorities
- Operating department leader provides coaching to line leaders

	Focus	Who	Information Needed
Safety	Incident	Line Leaders	Incident forms
	Unsafe condition & behaviour	Line Leaders	Verbal information
	Behavior Observation Completion	Line Leaders	Verbal information
Quality	Quality Issues	Line Leaders	Incident forms & sample
	Parts Per Million Defects	Line Leaders	Output Tracking
	Total Scrap	Line Leaders	Output Tracking
Throughput	Unplanned stop (#)	Line Leaders	Output Tracking
	Unplanned PR Loss (%)	Line Leaders	Output Tracking
	Previous Shift Production	Line Leaders	Output Tracking
	Planned Production Planned Upcoming Shift	Line Leaders	Output Tracking
	Changer Over (C/O) plan and time	Line Leaders	Output Tracking
Stability	Clean Inspect and Lubricate (CIL) plan	Line Leaders	CIL history
	Current Team (CIL) Completion	Line Leaders	CIL history
	False Starts	Line Leaders	False Starts Report

Participants
1. Operatging Department Leader   3. Process Engineers   5. Finished Product Handling Leader
2. Line Leaders   4. Material Supply Leader

Safety is the first item to address. The person responsible for Safety leads this session. This person asks each line leader for comment. Use of red, green and yellow cards allows for quick indication of the safety status. Only those showing a yellow or red card need to talk.

The person responsible for Quality follows a similar process.

The person responsible for identifying throughput issues follows as similar proccss.

Throughput is addressed by the department manager. The department manager announces the expected product production by line and highlights any special line action that will be required.

Again, only the line that has a problem signals in red. The rest show a green.

The logic to get resources for a specific problem focuses first on the line operating team solving their own problem. If the team does not have the skill or have other issues that prevent them from solving the problem, the resource request follows the logic.

Output Tracking Used to ID Resource Need

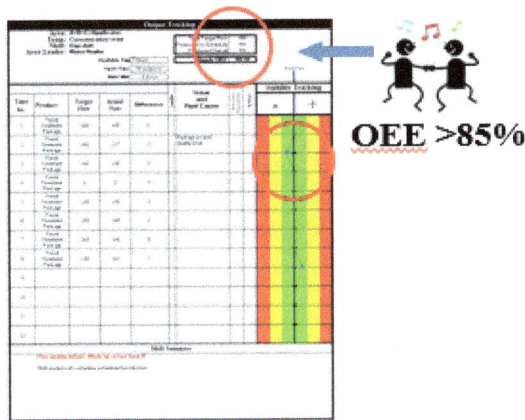

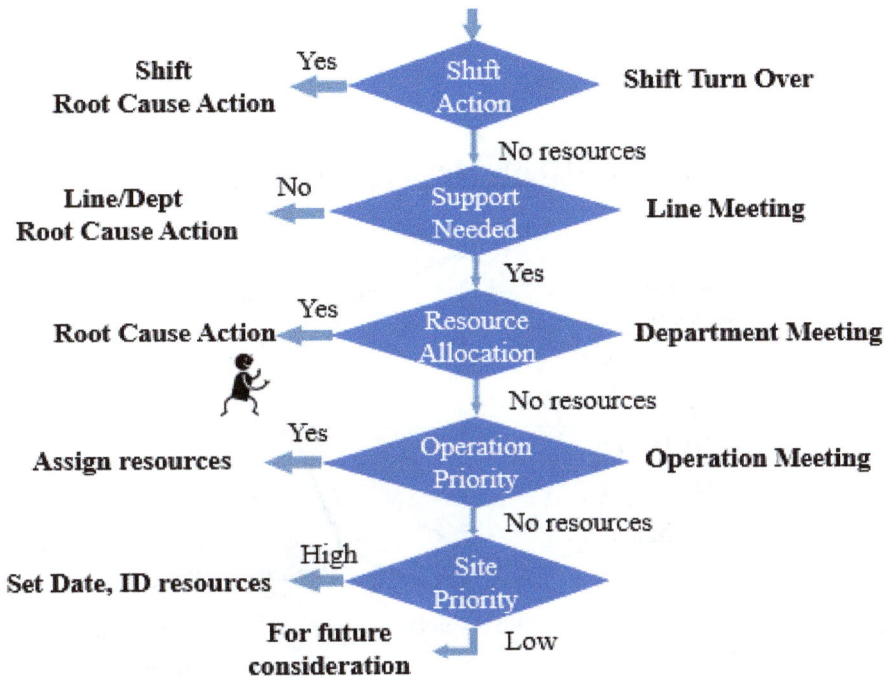

OEE >85%

Shift Root Cause Action	← Yes	**Shift Action** → **Shift Turn Over**
		No resources ↓
Line/Dept Root Cause Action	← No	**Support Needed** → **Line Meeting**
		Yes ↓
Root Cause Action	← Yes	**Resource Allocation** → **Department Meeting**
		No resources ↓
Assign resources	← Yes	**Operation Priority** → **Operation Meeting**
		No resources ↓
Set Date, ID resources	← High	**Site Priority**
For future consideration		↓ Low

This resource need logic flows up to the site priority level.

Individual, Team and Department Skill Evaluation

Individual Skill Summary

Equipment and people do work. It takes maintenance to keep equipment in good condition. Equipment needs constant maintenance. It never improves.

People do work and if nurtured grow and continuously improve. They take care of it and improve it.

Monitoring the skills of every individual and helping them set skill improvement goals ensures that the individual is improving in the currently needed skills. It also allows setting of skill growth that will be needed in the future.

Using a visual display of the individual's skills provides a way for leadership to coach the individual.

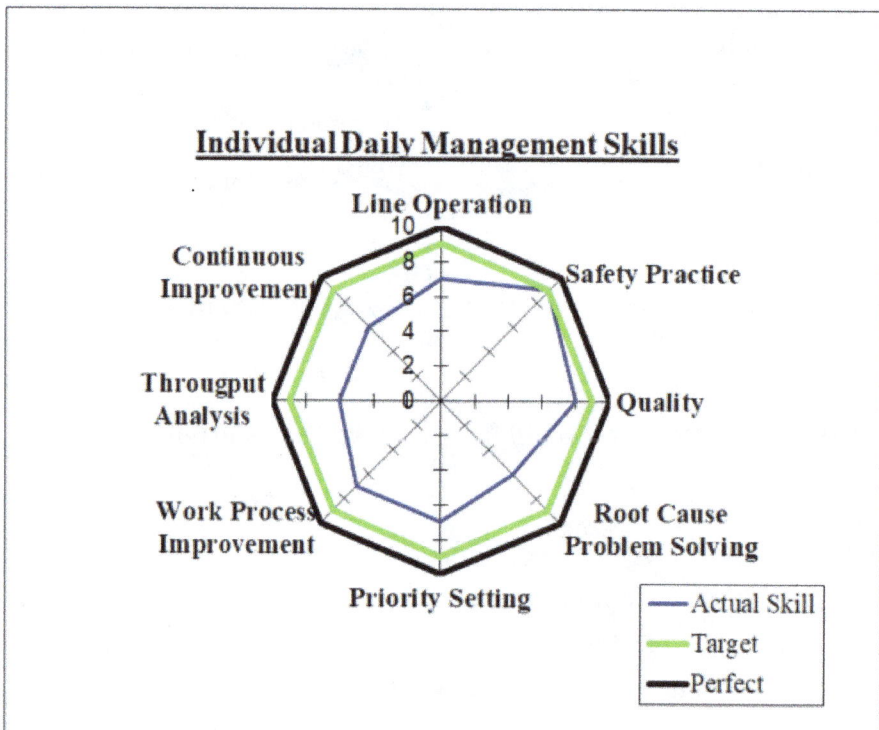

Individual Daily Management Skills

Line Operation

Continuous Improvement

Throught Analysis

Work Process Improvement

Priority Setting

Safety Practice

Quality

Root Cause Problem Solving

—Actual Skill
—Target
—Perfect

Line Team Skill Summary

A visual line team, composite skill summary is a good way for line leader and department manager to understand what skills need to be further developed in the operating team. It also allows line team members to select the skill growth that best helps the team.

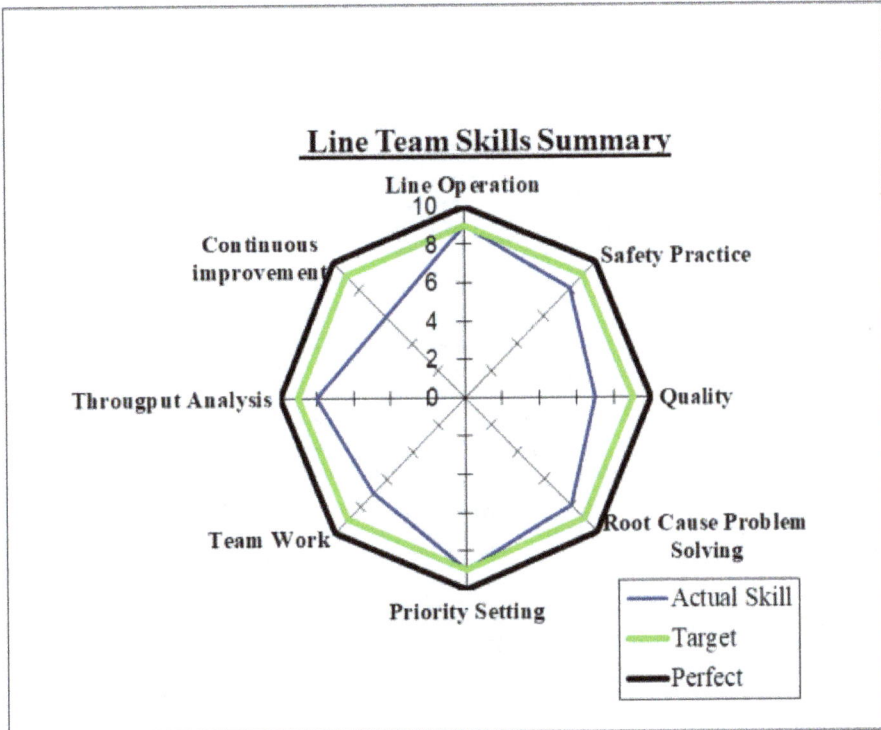

The lowest score for this team is in their continuous improvement area. However, they could also improve in how they handle quality issues and they also have an opportunity in improving their teamwork.

Department Skill Summary

The composite skill profiled for the department guides the department manager in choosing the areas to focus the coaching and provides guidance in the selection of the training to give the department member

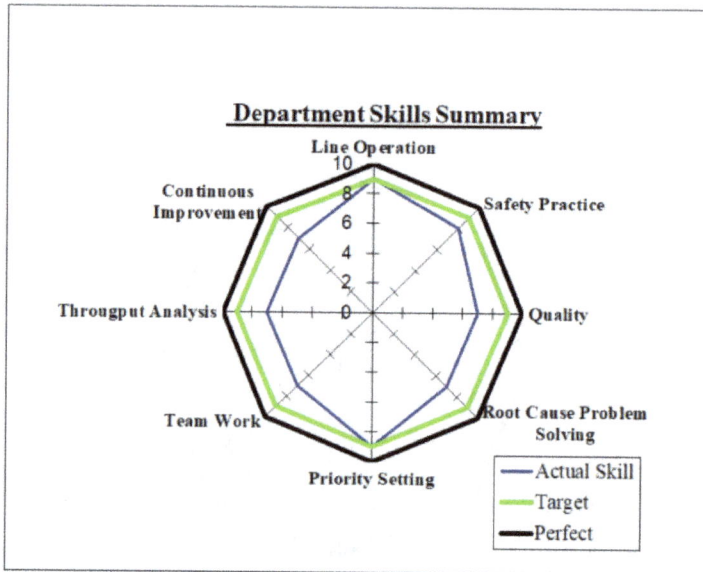

Department Skills Summary

This diagram would guide me to select root cause problem solving training for the department. Root Cause training would most likely improve the throughput analysis capability and improvements coming out of solving some problems at root cause may improve the quality rating.

Products of Chapter 3:
1. Aligned, focused department
2. Safe Operation
3. Stability and Product produced at Quality
4. Zero Loss Next twenty-four hours.

Chapter 3 Tips and Tools
1. Use Standard Meeting Agenda
2. Use output tracking
3. Use the resource allocation logic.
4. Use Next Twenty-Four Task List
5. Schedule Support help

Chapter 4: Operation Daily Management

The daily management process began with the operating line team and its leader. The line leader met with the line support team to discuss needs of the line team and to understand support actions that might impact the line operation. It then brought together all the operating team leaders in a meeting with the department leader. The department leader gained a detailed understanding of the line needs and then proceeded on to the Operations daily meeting.

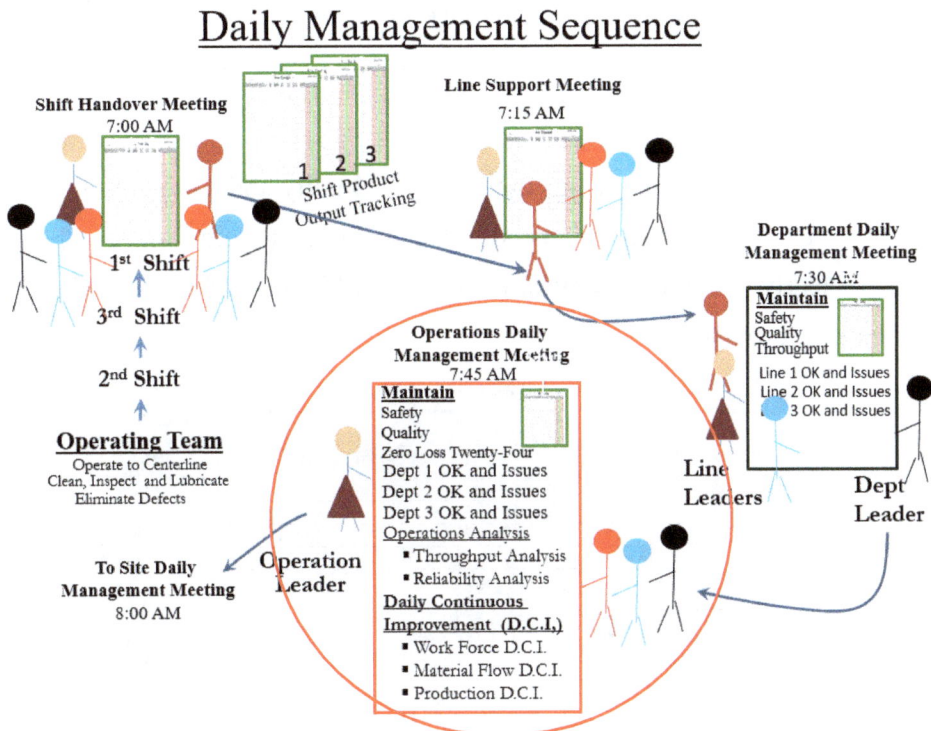

Daily Management Sequence

The Operations Daily Management Meeting has two parts; the **Maintain** and **Continuous Improvement**.

Maintain

The Maintain part of the agenda is a duplicate of all previous agendas. The maintain items are addressed by each department and support area as needed.

	Operations Meeting		
OBJECTIVE: To maitain next 24 hour production across the departments To review the progress of the continuous improvements			
	Focus	**Who**	**Information Needed**
	Safety	Operating Department Managers	Safety incident forms
	Quality	Quality Leaders	Quality incident forms
	Throughput Analysis	Operating Department Managers	From Output Tracking form
	Production Plan	Finished product planners	Production Plan
	Critical SKUs and qualifications	Finished product planners	Production Plan
	Stop Plan for next 24 hrs	Operating Department Managers	Production Plan
Maintain	Production Line Constraints	Operating Department Managers	Verbal information
	Through Put Analysis	Operating Department Managers	From Output Tracking form
	Reliability Analysis	Operating Department Managers	From Output Tracking form
	OEE Review	Operating Department Managers	From Output Tracking form
	Inventory Performance	Warehouse leader	From WHS. Output Tracking form
	Delivery	Operating Department Managers	Output Tracking Summary
Continuous Improvement	Operation Organiztion Skill Spider Diagram Evauation	Operating Department Managers	Line and Composite Spider Diagrams
	Work Force Continuous Improvement	Operating Department Managers	Specifics based on Improvement
	Material Flow Continuous Improvement	Operating Department Managers	Specifics based on Improvement
	Production Continuous Improvement	Operating Department Managers	Specifics based on Improvement
	Environment Continuous Improvement	Operating Department Managers	Specifics based on Improvement

PARTICIPANTS
1. Operation Manager   5. Warehouse leader *   *Invited based on situation
2. Department Managers   6. Quality leader*
3. Raw Materials Planners*   7. Material supply leader*
4. Finished product planners *

Each department leader brings their Safety, Quality and Throughput priority needs. These needs are visually displayed in a Pareto format. The goal is to clearly communicate what resources will be needed to maintain continuous improvement.

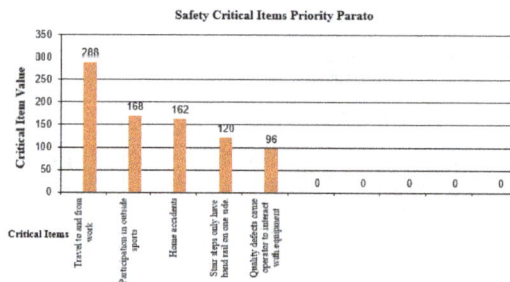

Safety Critical Items Priority Parato

Quality Critical Items Priority Parato

Throughput Critical Items Priority Parato

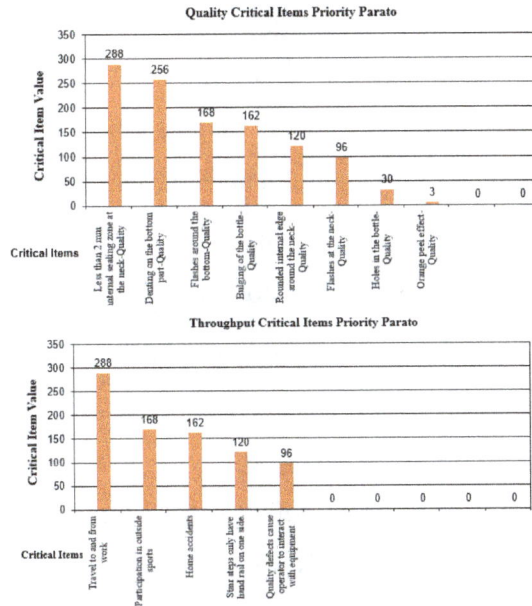

Continuous Improvement

The continuous improvement items are briefly discussed.

Continuous Improvement			
	Work Force Continuous Improvement	Operating Department Managers	Specifics based on Improvement
	Material Flow Continuous Improvement	Operating Department Managers	Specifics based on Improvement
	Production Continuous Improvement	Operating Department Managers	Specifics based on Improvement
	Environment Continuous Improvement	Operating Department Managers	Specifics based on Improvement

Work Force Continuous Improvement

Skill and capability assessment of individuals and teams are used to determine the training support required to ensure the entire organization is improving and prepared for new organization challenges. The goal is to develop and organization capable dealing with the change that technology and the marketplace creates.

Material Flow Continuous Improvement

Maintaining the minimum inventory of all raw materials but having them available for all production volume cycles is the goal of material flow improvement work. Output tracking and quantity flow control provides the basis for maintaining a minimum material required for production.

Production Continuous Improvement

There are small improvements such as reducing product to product changeover time, to larger improvements such as increasing production throughput on two lines so a third production line can be retired. In most production systems there are a large number of small but significant improvements. The big improvements normally present themselves when the product is being improved and in turn causes the production system to be changed.

Environment Continuous Improvement

Improved lighting, a leak proof roof, a better floor skid proof coating, improvement in the break area, an improved lunch area, are all examples of environment continuous improvement. These improvements impact the individuals in the work area. They may impact all the maintain metrics and in some cases they make the work area more pleasing.

Conclusion

The goal is for daily progress to occur. Each department leads on one item and if the improvement they lead is successful the other departments are responsible for reapplication,

Upon completion of the Operations Daily meeting the operations leader moves on to the site leadership daily meeting.

Products of Chapter 4
1. Aligned, focused operating department
2. Safe Operation
3. Stability and Product produced at Quality
4. Zero Loss Next twenty-four hours.
5. Continuous Improvement

Chapter 4 Tips and Tools
1. Use Standard Meeting Agenda
2. Use output tracking
3. Use of Priority Matrixes
4. Use Next Twenty-Four Task List
5. Schedule Support help

Chapter 5: Site Daily Management Meeting

There are two forms for the site leadership meeting. The one described in this chapter is referred to as the common version. The one described in Chapter 7: Daily Management Leadership Walk is more interactive and on the floor approach. It requires close coordination among all the leaders.

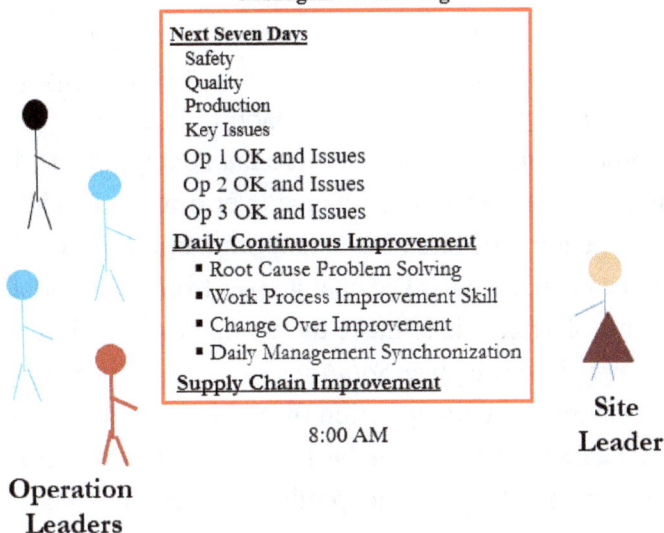

Site Leadership
Daily Management Meeting

Site Daily Management Meeting

Next Seven Days
Safety
Quality
Production
Key Issues
Op 1 OK and Issues
Op 2 OK and Issues
Op 3 OK and Issues
Daily Continuous Improvement
- Root Cause Problem Solving
- Work Process Improvement Skill
- Change Over Improvement
- Daily Management Synchronization
Supply Chain Improvement

8:00 AM

Operation Leaders

Site Leader

The site daily management meeting agenda has three parts.

	Focus	Who	Information Needed
	Site Leader's Meeting		
	OBJECTIVE: To prioritise support across all operations. To ensure clear site wide priority is clear. To ensure continuous improvement is executed. To understand the supply chain barriers and improvement opportunities		
Next Seven Days	Safety	Site Safety Owner	Site Safety Issues Summary
	Quality	Site Quality Owner	Site Quality Issues Summary
	Production	Site Production Plan Leader	Site Output Tracking Summary
	Key Issues	Site Leader	Key Issues List
Continuous Improvement	Root Cause Problem Solving	Problem Solving Owner	Individual Skill Cards
	Work Process Improvment Skill	Work Process Improvement Owner	Individual Skill Cards
	Change Over Monitoring	Operations Managers	Change Over Monitoring Process
	Daily Management Synchronization	Operations Managers	Daily Mgmt Monitoring Process
SC	Supply Chain Improvement	Site Manager	Boundary Output Tracking
	PARTICIPANTS 1. Site Leader 2. Operation managers 3. Support organization leaders		

Part One

The first part is similar to all the agendas coming up from the line and addresses remaining issues for Safety, Quality, Throughput, and any issues still requiring resources. This is the Maintain part of the agenda.

Part Two

The second part of the agenda focuses on Daily Continuous Improvement:

- Root cause problem solving is a critical skill for every individual. *Stress Free*[TM] *Manufacturing Solutions* addresses this capability in detail.
- Work process improvement is the next item that provides the organization a way to stabilize the work activities. *Stress Free*[TM] *Work Process Solutions* addresses this capability in detail.
- Many production lines produce more than one product. The ultimate goal is to have a push button change over from on product to the next. However, a common situation is that production line equipment parts need to be changed out. The changeover from one product to the next costs time. The goal is to make the change over as short as possible. *Stress Free*[TM] *Changeover Solutions* addresses this capability in detail.
- Synchronizing Leadership action from the production line up through the Site Leadership forms the basis of a production operation that performs in an optimum, safe, productive way. Getting the organization leaders synchronize is covered in this book; *Stress Free*[TM] *Daily Management Solutions*

Part Three

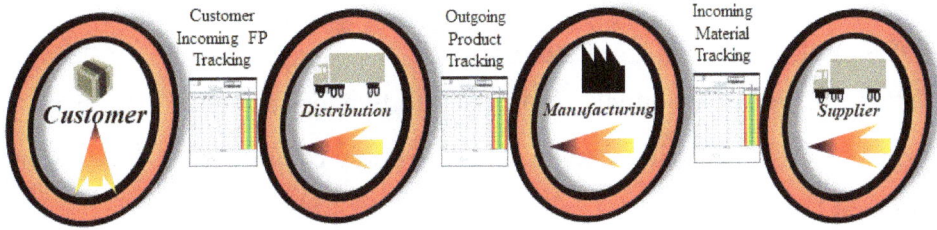

The final part of the agenda focuses on the site supply chains. The operations leaders and the site leaders discuss the issues from the suppliers of materials, the manufacturing issues affecting flow, the distribution issues and the flow of finished products to the customers.

The customer is the primary focus, but productivity, quality and cost all are important.

Note that the Output Tracking sheet is used between each of the supply chain interfaces to identify the critical issues and losses.

Outside the site participants

On a monthly basis and a rolling three-month focus, the supply chain leaders participate with other participating supply chain members. The goal is to address supply chain boundary issues and prioritize problem resolution.

Supply Chain Meeting			
OBJECTIVE: To evaluate category performance To identify common category problems			
	Focus	**Who**	**Information Needed**
Quaterly Performance	Safety Performance	Site Safety Leader	Site Safety Report
	Quality Performance	Site QualityLeader	Site Quality Report
	Production Performance	Site Production Leader	Production Report by SKU
	Inventory Performace	Site Production Planning Leader	Site Inventory Report by SKU
	Supply Chain Interface Issues	Supply Chain Partners	Supply Chain Output Tracking Sheets
Next Quarter focus	Safety Performance	Site Safety Leader	Category Safety Goal
	Quality Performance	Site QualityLeader	Category Quality Goal
	Production Performance	Site Production Leader	Production Performance Goal
	Inventory Performace	Site Production Planning Leader	Category Inventory Goal
	Supply Chain Interface Issues	Supply Chain Partners	Supply Chain Issues List

PARTICIPANTS
1. Category Leader
2. Site Leaders
3. Category Support Organization leaders
4. Participating Suppliers
5. Participating Customers
6. Participating Transport and logistic partners

Periodically the participants on the leadership team utilize a zero-loss assessment to better understand the supply chain situation.

Organization Zero Loss Assessment

The losses along the supply chain fall into five loss categories; Human Effort Losses, Organization Design Losses, Supply Chain Losses, Process and Equipment Losses and Other Losses.

Organization Zero Loss Assessment

Human Effort Losses

Human effort losses have two elements

1. **Leadership Losses**

   Leadership losses consider having a zero-loss vision, a master plan with key strategies and objectives. It considers the leaders actions, such as building trust and staffing for excellence,

2. **Personal Mastery Losses**

   Personal mastery losses consider that the organization has a goal to have every individual achieve mastery. It considers if the skills to maintain the business at zero loss have been chosen and that the needs of the business have been integrated into each individual's skill development plan.

   It also considers if education and training effectiveness is measured and if losses due to lack of skill are measured and trending down.

Organization Design Losses

Organization design losses have two loss elements

1. **Organizational Culture**
   It checks to see if there are documented principles, strategies and objectives that are clear to all employees.
   Employees are asked if they can link their efforts to the business goals.
   It checks to see that the organization encourages and assists individuals to develop their skills.
   It checks whether effective, capable teams exist to address interdependent work.

2. **Organizational Design**
   This area looks at recruiting practices, team oriented work design, flexible resource allocation, the reward and recognition system and if proper staffing is the norm for most work areas.

Supply Chain Losses

Five supply chain loss areas are considered

1. **Incoming materials logistics**
   Incoming raw materials and parts losses are reviewed.
   Incoming raw materials and parts quality validation are reviewed.
   The number of times each raw material or part is handled is checked.
   The material delivery timeliness and quantity to the production line is evaluated.
   The material and parts handling damage is considered.

2. **Outbound finished product logistics**
   Outbound finished product handling losses, grouping losses, staging losses, truck loading losses and customer receiving losses are all evaluated.

3. **Production area losses**
   Line startup losses, material and quality losses, line operational losses, line shutdown losses and line maintenance losses are evaluated.

4. **Yield losses**
   Material use losses, remnant losses, minor stop losses, destructive quality testing losses and finished product handling losses are all considered in this category.

5. **Production flow losses**
   Raw material flow, material handling effort losses, flow maintenance, finished product and handling flows are examined.

Process and Equipment Losses

Process and Equipment Losses

Three elements are examined in this category

1. **Equipment down time losses**
Maintenance losses, equipment setup, startup and failure are evaluated.
Waiting for materials of product tests losses are evaluated.

2. **Equipment performance losses**
Minor stops, rate variability, equipment idling, reduced speed operation and safety related losses are evaluated.

3. **Quality defect losses**
Raw material defect losses, product defect losses, quality testing, product rework and product handling quality losses are part of this area's evaluation.

Other Losses

Other Losses

Production Scheduling losses, energy related losses, utilization of equipment and facilities losses, employee health, housing and transportation to work losses and environmental losses are under the other category.

Supply Chain Critical Item Task List

The supply chain losses have been reviewed. The supply chain participants now have an understanding about the condition of the supply chain and having used output and input tracking at each supply chain boundary they are prepared to list and rank the critical items.

This activity provides a way for the participants to discuss the issues that affect each of them.

The rating and the subsequent selection of the improvement action to take provides focus.

Supply Chain Critical Item Task List

	Critical Item	Late	Quality Issue	Quantity Issue	Criticality Value
			One to Ten Rating		
1	Bottle Label Supplier	3	7	5	105
2	Bottle Supplier	7	3	7	147
3	Cap Supplier	6	3	9	162
4	Truck arrival	5	4	6	120
5	Material to production line delviery	6	3	7	126
6	Product to customer dock	7	4	6	168
7					0
8					0
9					0
10					0

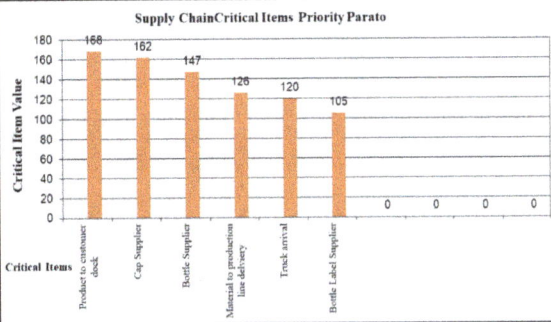

Supply ChainCritical Items Priority Parato

Rating	Priority
1=	Low Priority
8=	Low Priority
27=	Low Priority
64=	Low Priority
125=	Medium
343=	Medium
512=	High
729=	Very High
1000=	Critical

Prioritize

Item	Priority	Resolution Owner
Product to customer dock	168	
Cap Supplier	162	
Bottle Supplier	147	
Material to production line delviery	126	
Truck arrival	120	
Bottle Label Supplier	105	
	0	
	0	
	0	
	0	

This process is one of the more useful supply chain activities.

Products of Chapter 5

1. Aligned, focused Site
2. Aligned focused supply chain leaders
3. Supply Chain Stability

Chapter 5 Tips and Tools

1. Use Output and Input Tracking sheet
2. Use Organization and Supply Chain Loss Assessment
3. Use Critical Items Task List

Chapter 6: Support Area Daily Management

The Support Areas are:

- The Safety Department
- The Quality Department
- The Engineering Department
- The Finance Department
- The Office Support group.

Each of these organizations hold a morning meeting. Because the people in these organizations participate in the production daily management meetings, the timing for these meetings follow immediately after the line daily management meetings.

Quality Daily Management Flow

All support areas practice a similar daily management flow.

This is an example of the Quality department daily meeting. The Quality Department focuses on; current production line issues, finished product quality issues and raw material quality issues,

		Focus	Who	Information Needed
Finished Product		Safety as related to Product Quality	Site Safety Owner	Site Safety Issues Summary
		Product Quality	Site Finished Product Quality Owner	Site Product Quality Issues Summary
		Finished Product PPM Defect Rate	Site Finished Product Quality Owner	Product Defect Tracking List
		Key Finished Product Quality Issues	Site Finished Product Quality Owner	KeyFinished Product Qulity Issues List
Raw Materials		Safety as related to Materials Quality	Site Safety Owner	Site Safety Issues Summary
		Materials Quality	Site Materials Quality Owner	Site Materials Quality Issues Summary
		Materials PPM Defect Rate	Site Materials Quality Owner	Product Defect Tracking List
		Key Materials Quality Issues	Site Quality Owner	Key Materials Quality Issues List
Critical Items		Critical Item 1	Site Quality Owner	Quality Critical Items List
		Critical Item 2	Site Quality Owner	Quality Critical Items List

Quality Support Area Daily Meeting

__OBJECTIVE__: To prioritise Quality support across all operations.
To ensure Zero Losses due to Quality

PARTICIPANTS
1. Site Quality Leader   4. Department Representatives*   * Attendance as needed
2. Product Quality Owners   5. Site Safety Owner*
3. Materials Quality Owners

Current Production Quality Issues

The previous shift and the upcoming production shift quality issues are discussed. Support for getting the production to 100% quality production is discussed and resources are assigned to support the problem area.

Finished Product Quality Issues

The finished product from each line is tested and evaluated. Additionally, the outgoing product is tested to assure no handling and transport damage is occurring.

Raw Materials Quality Issues

Each raw material is checked to ensure it meets the quality criteria and has no damage. This checking is done by the person receiving the material. The quality may have been certified by the material provider or it may undergo random quality checks upon receipt. The goal is to have defect free material flowing to the production line.

Quality Critical Items List

This is the overall list of critical quality issues whether raw, finished product quality or quality related safety issues.

Quality Critical Item List

Critical Item	Safety	Quality	Throughput	Criticality Value
Flashes at the neck- Quality	4	3	8	96
Rounded internal edge around the neck- Quality	3	8	5	120
Less than 2 mm internal sealing zone at the neck-Quality	4	8	9	288
Bulging of the bottle-Quality	9	6	3	162
Flashes around the bottom-Quality	4	6	7	168
Orange peel effect-Quality	3	1	1	3
Holes in the bottle-Quality	3	5	2	30
Denting on the bottom part-Quality	8	8	4	256
				0
				0

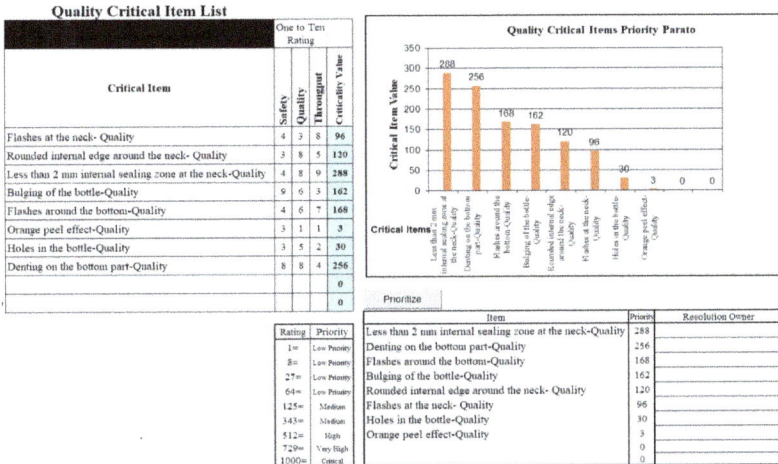

Quality Critical Items Priority Parato

Rating	Priority
1=	Low Priority
8=	Low Priority
27=	Low Priority
64=	Low Priority
125=	Medium
343=	Medium
512=	High
729=	Very High
1000=	Critical

Prioritize

Item	Priority	Resolution Owner
Less than 2 mm internal sealing zone at the neck-Quality	288	
Denting on the bottom part-Quality	256	
Flashes around the bottom-Quality	168	
Bulging of the bottle-Quality	162	
Rounded internal edge around the neck- Quality	120	
Flashes at the neck- Quality	96	
Holes in the bottle-Quality	30	
Orange peel effect-Quality	3	
	0	
	0	

Issues are listed and then rated as to impact on Safety, Quality and Throughput on a one to five scale, where five is the most critical. The priority is visualized with a graph and the items can be assigned to a resolution owner.

The rating value is compared to a rate and priority scale to better understand the how critical an item may be.

All issues are important but an issue that is rated high or above must receive priority attention.

Products of Chapter 6

1. Aligned, focused support areas
2. Critical Items identified

Chapter 6 Tips and Tools

1. Use Standard Meeting Agenda
2. Use output tracking
3. Use of Priority Matrixes
4. Use Next Twenty-Four Task List
5. Set support help Schedule

Chapter 7: Leadership Daily Management Walk

The daily leadership walk is a dynamic on the floor way for leadership to interact at each level of the organization. It allows leadership at every level to interact with each other and it affords top leaders to interact and coach at every level.

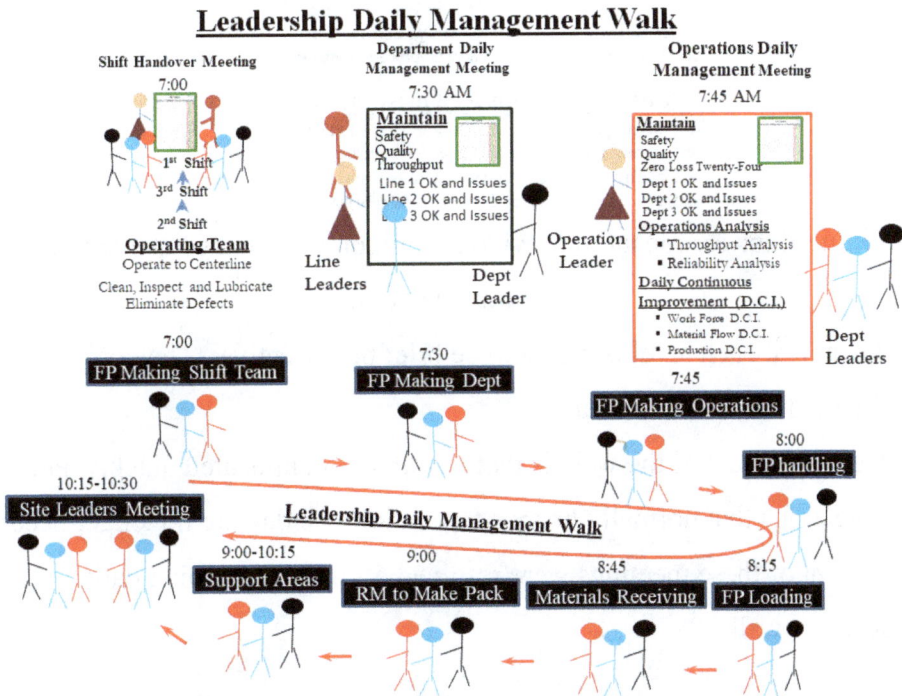

Leadership Daily Management Walk

The walk begins at the line team level. Since there are many line teams, a schedule is set so that every line team is visited during one complete leadership walk cycle. That logic is applied to every level of the organization,

The site leader or a designated replacement leads the team. This team is met by the leader for the area being visited. They listen to the area's daily meeting. They have a predetermined standard set of questions for each area.

Synchronized Daily Management Walk Support Questions

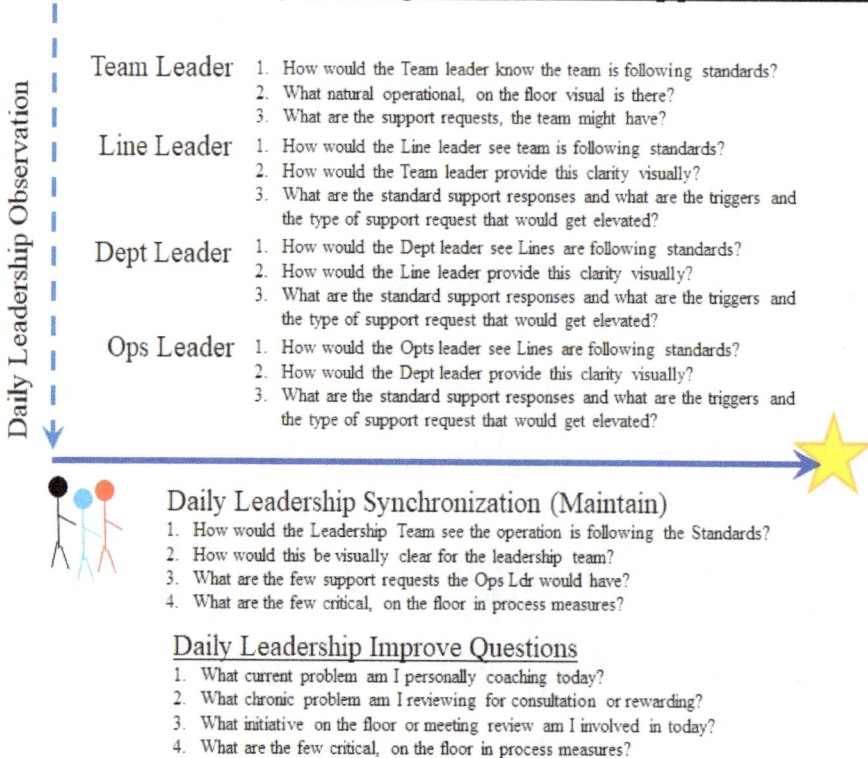

Team Leader
1. How would the Team leader know the team is following standards?
2. What natural operational, on the floor visual is there?
3. What are the support requests, the team might have?

Line Leader
1. How would the Line leader see team is following standards?
2. How would the Team leader provide this clarity visually?
3. What are the standard support responses and what are the triggers and the type of support request that would get elevated?

Dept Leader
1. How would the Dept leader see Lines are following standards?
2. How would the Line leader provide this clarity visually?
3. What are the standard support responses and what are the triggers and the type of support request that would get elevated?

Ops Leader
1. How would the Opts leader see Lines are following standards?
2. How would the Dept leader provide this clarity visually?
3. What are the standard support responses and what are the triggers and the type of support request that would get elevated?

Daily Leadership Observation

Daily Leadership Synchronization (Maintain)
1. How would the Leadership Team see the operation is following the Standards?
2. How would this be visually clear for the leadership team?
3. What are the few support requests the Ops Ldr would have?
4. What are the few critical, on the floor in process measures?

Daily Leadership Improve Questions
1. What current problem am I personally coaching today?
2. What chronic problem am I reviewing for consultation or rewarding?
3. What initiative on the floor or meeting review am I involved in today?
4. What are the few critical, on the floor in process measures?

These questions are meant to provide a common interchange at every level of the organization.

The stop at each level is meant to be brief but effective. The goal is to be done by the first half of the day.

One key impact of the walk is that resources get allocated quickly and many meetings that normally happen throughout the day are no longer needed. The content of these meetings were resolved during the morning daily management walk.

The walk path is designed to have a representative mix of leaders visit each level of the organization. The support area members are always on each walk. This provides a firsthand exposure of problems to the support area.

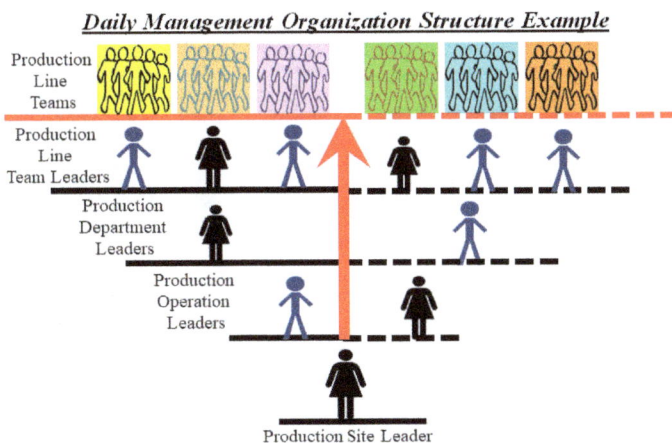

Daily Management Organization Structure Example

The leader at each level guides and manages the walking visitors. This approach gives each level leader exposure to the leadership team.

The entire organization benefits from this on the floor style of daily management,

Products of Chapter 7

1. Aligned, focused operating department
2. Safe Operation
3. Direct on the floor coaching
4. Zero Loss Next twenty-four hours.
5. Continuous Improvement

Chapter 7 Tips and Tools

1. Examine the use of Standard Meeting Agenda
2. Examine the use of output tracking
3. Examine the use of Priority Matrixes
4. Examine the use of Next Twenty-Four Task List

Chapter 8: The Building Blocks of Daily Management

Daily Morning Meeting Agenda's

Using a common set meeting agenda from the line to the site level provides a basis for a common interaction at every level of the organization

Shift Hand Over Meeting

The shift hand over meeting is the first of the series. The four main topics are Safety, Quality, Throughput and Stability

Shift Hand Over meeting			
OBJECTIVE: Understand gaps from previous shift to support setting the right priorities for the next shift. Peer to peer information handover to get basic understanding of the previous shift			
	Focus	**Who**	**Information Needed**
Safety	Incident	Shift line leader	Incident forms
	Unsafe condition & behaviour	Shift line leader	Verbal information
	Behavior Observation Plan	Shift line leader	-
Quality	Quality Issues	Shift quality operator owner	Incident forms & sample
	Finished product sample for oncoming team to check	Shift quality operator owner	one bag of product
Throughput	Planned production	Shift line leader	Upcoming shift production plan
	Production output	Shift line leader	Previous shift output tracking
	Top Loss - Unplanned Stops	Shift line leader	Previous shift output tracking
	Changer Over (C/O) plan	Shift line leader	Upcoming shift production plan
Stability	Clean Inspect and Lubricate (CIL) plan	Shift CIL operator owner	CIL history
	Current Team CIL Completion	Shift CIL operator owner	CIL history
	Equipment touches (ET)	Shift (ET) operator owner	Equipment Touches history and goal
	Defect elimination (DE)	Shift (DE) operator owner	Defect elimination history and goal
	Support need	Shift line leader	
PARTICIPANTS 1. On coming shift line leader 5. Shift Process Technician 2. Off going shift line leader 6. Shift maintenance technician 3. On coming line operators 7. Shift E&I technician			

The topic is discussed in detail only if there is an issue. If there are no issues for a topic a green card is held up or "green" is declared.

Line Support Meeting

This agenda is used to communicate between the line leader and the line support personnel. The intent is to make sure that the line has the required support for the next upcoming shift.

Line Support Meeting		

OBJECTIVE: - Review last 24 hrs results to identify top 3 losses, determine immediate actions and root cause actions.
- Determine and additional actions for shift team.
- Verify that the support resources are able to address any outstanding issues

	Focus	Who	Information Needed
Safety	Incident	Line Leader	Incident forms
	Unsafe condition & behaviour	Line Leader	Verbal information
	Behavior Observation Completion	Line Leader	Verbal information
Quality	Quality Issues	Line Leader	Incident forms & sample
	Parts Per Million Defects	Line Leader	Output Tracking
	Total Scrap	Line Leader	Output Tracking
Throughput	Unplanned stop (#)	Line Leader	Output Tracking
	Unplanned PR Loss (%)	Line Leader	Output Tracking
	Previous Shift Production	Line Leader	Output Tracking
	Production Planned for the upcoming shift	Line Leader	Output Tracking
	Changer Over (C/O) plan and time	Line Leader	Output Tracking
Stability	Clean Inspect and Lubricate (CIL) plan	Line Leader	CIL history
	Current Team (CIL) Completion	Line Leader	CIL history
	False Starts	Line Leader	False Starts Report
	Maintenance Requirements	Line Leader	Maintenance Schedule

PARTICIPANTS

1. Line Leader
2. Maintenance Technician
3. E&I Technician
4. Process Technician
5. Department Safety Owner (as needed)
6. Department Quality Owner (as needed)

Department Meeting Agenda

The department meeting provides an across the production lines view of the condition and any conflicts in support needs.

Department Daily Meeting Agenda

OBJECTIVE: - To align on departmental priorities
  - Operating department leader provides coaching to line leaders

	Focus	Who	Information Needed
Safety	Incident	Line Leaders	Incident forms
	Unsafe condition & behaviour	Line Leaders	Verbal information
	Behavior Observation Completion	Line Leaders	Verbal information
Quality	Quality Issues	Line Leaders	Incident forms & sample
	Parts Per Million Defects	Line Leaders	Output Tracking
	Total Scrap	Line Leaders	Output Tracking
Throghput	Unplanned stop (#)	Line Leaders	Output Tracking
	Unplanned PR Loss (%)	Line Leaders	Output Tracking
	Previous Shift Production	Line Leaders	Output Tracking
	Planned Production Planned Upcoming Shift	Line Leaders	Output Tracking
	Changer Over (C/O) plan and time	Line Leaders	Output Tracking
Stability	Clean Inspect and Lubricate (CIL) plan	Line Leaders	CIL history
	Current Team (CIL) Completion	Line Leaders	CIL history
	False Starts	Line Leaders	False Starts Report

Participants
  1. Operatging Department Leader   3. Process Engineers   5. Finished Product Handling Leader
  2. Line Leaders   4. Material Supply Leader

Resource conflicts are resolved. If there is a resource shortage or issue the department leader will take it to the Operations level meeting,

If there are specific problems that require outside support then the pertinent people are asked to attend the meeting.

Operations Meeting Agenda

The operations meeting brings together all the department managers and all support area leaders as needed.

Operations Meeting			
OBJECTIVE: To maitain next 24 hour production across the departments To review the progress of the continuous improvements			
	Focus	**Who**	**Information Needed**
	Safety	Operating Department Managers	Safety incident forms
	Quality	Quality Leaders	Quality incident forms
	Throughput Analysis	Operating Department Managers	From Output Tracking form
	Production Plan	Finished product planners	Production Plan
	Critical SKUs and qualifications	Finished product planners	Production Plan
	Stop Plan for next 24 hrs	Operating Department Managers	Production Plan
Maintain	Production Line Constraints	Operating Department Managers	Verbal information
	Through Put Analysis	Operating Department Managers	From Output Tracking form
	Reliability Analysis	Operating Department Managers	From Output Tracking form
	OEE Review	Operating Department Managers	From Output Tracking form
	Inventory Performance	Warehouse leader	From WHS. Output Tracking form
	Delivery	Operating Department Managers	Output Tracking Summary
Continuous Improvement	Operation Organiztion Skill Spider Diagram Evauation	Operating Department Managers	Line and Composite Spider Diagrams
	Work Force Continuous Improvement	Operating Department Managers	Specifics based on Improvement
	Material Flow Continuous Improvement	Operating Department Managers	Specifics based on Improvement
	Production Continuous Improvement	Operating Department Managers	Specifics based on Improvement
	Environment Continuous Improvement	Operating Department Managers	Specifics based on Improvement

PARTICIPANTS

1. Operation Manager
2. Department Managers
3. Raw Materials Planners*
4. Finished product planners *

5. Warehouse leader *
6. Quality leader*
7. Material supply leader*

*Invited based on situation

The agenda follows the pattern set up by the organizations below this level. The new items are those focused on continuous improvement. The department managers each have a specific improvement for which the take the lead. Once an improvement is made the other department managers are responsible for the reapplication of the improvement.

Site Meeting Agenda

The site daily meeting has a longer-term flavor. It looks at safety, quality, production, and other key issues with a seven-day forward look. The goal is to make sure that these three areas are in total control.

Site Leader's Meeting		
OBJECTIVE: To prioritise support across all operations. To ensure clear site wide priority is clear To ensure continuous improvement is executed To understand the supply chain barriers and improvement opportunities		

	Focus	Who	Information Needed
Next Seven Days	Safety	Site Safety Owner	Site Safety Issues Summary
	Quality	Site Quality Owner	Site Quality Issues Summary
	Production	Site Production Plan Leader	Site Output Tracking Summary
	Key Issues	Site Leader	Key Issues List
Continuous Improvement	Root Cause Problem Solving	Problem Solving Owner	Individual Skill Cards
	Work Process Improvment Skill	Work Process Improvement Owner	Individual Skill Cards
	Change Over Monitoring	Operations Managers	Change Over Monitoring Process
	Daily Management Synchronization	Operations Managers	Daily Mgmt Monitoring Process
SC	Supply Chain Improvement	Site Manager	Boundary Output Tracking

PARTICIPANTS
1. Site Leader
2. Operation managers
3. Support organization leaders

Continuous Improvement in Root Cause Problem Solving, Work Process Improvement, Change Over and Daily Management Synchronization are each address. The goal is to make steady progress. Each area has an owner at the top leadership level responsible in sponsoring and coaching the continuous improvement.

Supply Chain Improvement is the site manager's responsibility. There may be other members on the leadership team involved. The goal is to create a smoothly flowing stream of materials and finished product with only a three sigma inventory level.

Supply Chain Meeting Agenda

The supply Chain "daily" management meeting is usually scheduled once per month. It always looks back at the last three months and then looks forward for what issues to resolve.

Supply Chain Meeting			
OBJECTIVE: To evaluate category performance To identify common category problems			
	Focus	**Who**	**Information Needed**
Quaterly Performance	Safety Performance	Site Safety Leader	Site Safety Report
	Quality Performance	Site QualityLeader	Site Quality Report
	Production Performance	Site Production Leader	Production Report by SKU
	Inventory Performace	Site Production Planning Leader	Site Inventory Report by SKU
	Supply Chain Interface Issues	Supply Chain Partners	Supply Chain Output Tracking Sheets
Next Quarter focus	Safety Performance	Site Safety Leader	Category Safety Goal
	Quality Performance	Site QualityLeader	Category Quality Goal
	Production Performance	Site Production Leader	Production Performance Goal
	Inventory Performace	Site Production Planning Leader	Category Inventory Goal
	Supply Chain Interface Issues	Supply Chain Partners	Supply Chain Issues List
PARTICIPANTS			
1. Category Leader	4. Participating Suppliers		
2. Site Leaders	5. Participating Customers		
3. Category Support Organization leaders	6. Participating Transport and logistic partners		

The use of the output tracking sheet between each organizational boundary provides specific performance information allows the partnering organizations to address specific problems. The Supply Chain Issues list prioritizes the action that should be taken during the coming month. This approach makes all supply chain participants partners in creating a minimum loss, flowing supply chain,

Support Area Meeting Agendas

The support areas use a similar agenda being used throughout the organization. The support organization meeting takes place immediately after the line daily management meetings end. This allows them to have fresh information as to the condition the work they need to do in support of the organization.

	Quality Support Area Daily Meeting		
	OBJECTIVE: To prioritise Quality support across all operations. To ensure Zero Losses due to Quality		
	Focus	**Who**	**Information Needed**
Finished Product	Safety as related to Product Quality	Site Safety Owner	Site Safety Issues Summary
	Product Quality	Site Finished Product Quality Owner	Site Product Quality Issues Summary
	Finished Product PPM Defect Rate	Site Finished Product Quality Owner	Product Defect Tracking List
	Key Finished Product Quality Issues	Site Finished Product Quality Owner	KeyFinished Product Qulity Issues List
Raw Materials	Safety as related to Materials Quality	Site Safety Owner	Site Safety Issues Summary
	Materials Quality	Site Materials Quality Owner	Site Materials Quality Issues Summary
	Materials PPM Defect Rate	Site Materials Quality Owner	Product Defect Tracking List
	Key Materials Quality Issues	Site Quality Owner	Key Materials Quality Issues List
Critical Items	Critical Item 1	Site Quality Owner	Quality Critical Items List
	Critical Item 2	Site Quality Owner	Quality Critical Items List

PARTICIPANTS
1. Site Quality Leader
2. Product Quality Owners
3. Materials Quality Owners
4. Department Representatives*
5. Site Safety Owner*

* Attendance as needed

Critical Items Task List

The critical items task list provides a convenient way to visually display the priority issues. This for can be customized to fit specific organization or specific focus.

Critical Item Task List

Critical Item	Safety	Quality	Throughput	Criticality Value
Flashes at the neck- Quality	4	3	8	96
Rounded internal edge around the neck- Quality	3	8	5	120
Less than 2 mm internal sealing zone at the neck-Quality	4	8	9	288
Bulging of the bottle-Quality	9	6	3	162
Flashes around the bottom-Quality	4	6	7	168
Orange peel effect-Quality	3	1	1	3
Holes in the bottle-Quality	3	5	2	30
Denting on the bottom part-Quality	8	8	4	256
Impurities	10	8	7	560
				1

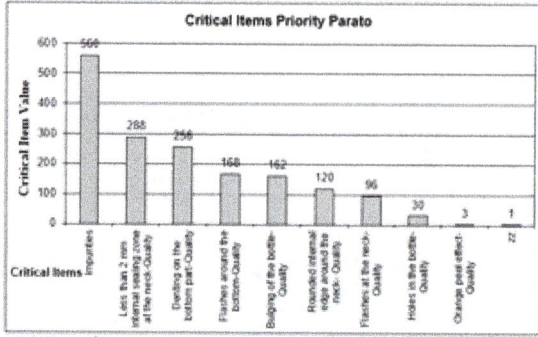

Critical Items Priority Parato

Rating	Priority
1=	Low Priority
8=	Low Priority
27=	Low Priority
64=	Low Priority
125=	Medium
343=	Medium
512=	High
729=	Very High
1000=	Critical

Prioritize

Item	Priority	Resolution Owner
Impurities	560	
Less than 2 mm internal sealing zone at the neck-Quality	288	
Denting on the bottom part-Quality	256	
Flashes around the bottom-Quality	168	
Bulging of the bottle-Quality	162	
Rounded internal edge around the neck- Quality	120	
Flashes at the neck- Quality	96	
Holes in the bottle-Quality	30	
Orange peel effect-Quality	3	
	1	

Safety Critical Items List

This is the critical items list as it might be applied to safety issues.

Quality Critical Item List

	Critical Item	Safety	Quality	Throughput	Criticality Value
1	Flashes at the neck- Quality	4	3	8	96
2	Rounded internal edge around the neck- Quality	3	8	5	120
3	Less than 2 mm internal sealing zone at the neck-Quality	4	8	9	288
4	Bulging of the bottle-Quality	9	6	3	162
5	Flashes around the bottom-Quality	4	6	7	168
6	Orange peel effect-Quality	3	1	1	3
7	Holes in the bottle-Quality	3	5	2	30
8	Denting on the bottom part-Quality	8	8	4	256
9					0
10					0

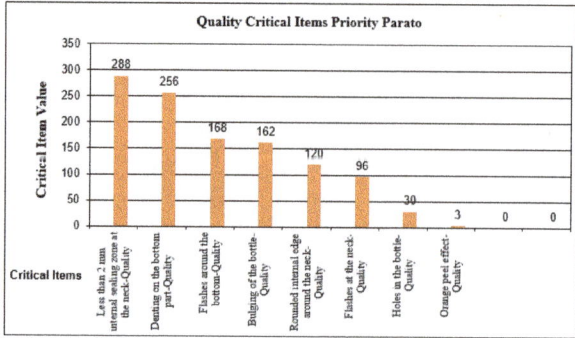

Quality Critical Items Priority Parato

Rating	Priority
1=	Low Priority
8=	Low Priority
27=	Low Priority
64=	Low Priority
125=	Medium
343=	Medium
512=	High
729=	Very High
1000=	Critical

Prioritize

Item	Priority	Resolution Owner
Less than 2 mm internal sealing zone at the neck-Quality	288	
Denting on the bottom part-Quality	256	
Flashes around the bottom-Quality	168	
Bulging of the bottle-Quality	162	
Rounded internal edge around the neck- Quality	120	
Flashes at the neck- Quality	96	
Holes in the bottle-Quality	30	
Orange peel effect-Quality	3	
	0	
	0	

Quality Critical Items List

This is the critical items list as it might be applied to Quality issues.

Quality Critical Item List

	Critical Item	One to Ten Rating			
		Safety	Quality	Throughput	Criticality Value
1	Flashes at the neck- Quality	4	3	8	96
2	Rounded internal edge around the neck- Quality	3	8	5	120
3	Less than 2 mm internal sealing zone at the neck-Quality	4	8	9	288
4	Bulging of the bottle-Quality	9	6	3	162
5	Flashes around the bottom-Quality	4	6	7	168
6	Orange peel effect-Quality	3	1	1	3
7	Holes in the bottle-Quality	3	5	2	30
8	Denting on the bottom part-Quality	8	8	4	256
9					0
10					0

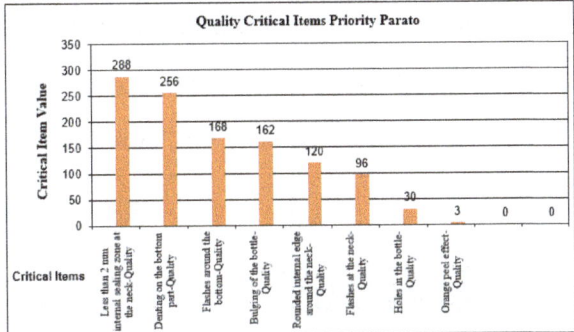
Quality Critical Items Priority Parato

Rating	Priority
1=	Low Priority
8=	Low Priority
27=	Low Priority
64=	Low Priority
125=	Medium
343=	Medium
512=	High
729=	Very High
1000=	Critical

Prioritize

Item	Priority	Resolution Owner
Less than 2 mm internal sealing zone at the neck-Quality	288	
Denting on the bottom part-Quality	256	
Flashes around the bottom-Quality	168	
Bulging of the bottle-Quality	162	
Rounded internal edge around the neck- Quality	120	
Flashes at the neck- Quality	96	
Holes in the bottle-Quality	30	
Orange peel effect-Quality	3	
	0	
	0	

Throughput Critical Items List

This is the critical items list as it might be applied to production throughput issues.

Throughput Critical Item Task List

	Critical Item	One to Ten Rating			
		Safety	Quality	Throughput	Criticality Value
1	Quality defects	4	5	8	160
2	Major Maintenance	3	8	5	120
3	Raw Material Quality	3	7	9	189
4	Case packer reliability	9	6	5	270
5					0
6					0
7					0
8					0
9					0
10					0

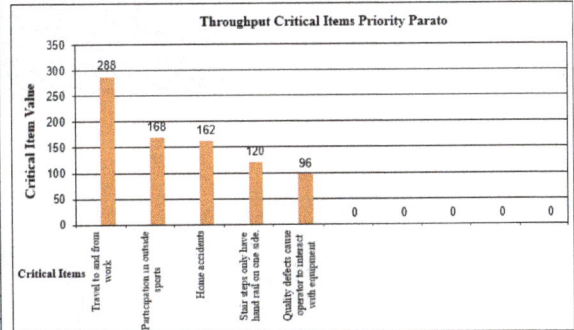
Throughput Critical Items Priority Parato

Rating	Priority
1=	Low Priority
8=	Low Priority
27=	Low Priority
64=	Low Priority
125=	Medium
343=	Medium
512=	High
729=	Very High
1000=	Critical

Prioritize

Item	Priority	Resolution Owner
Case packer reliability	270	
Raw Material Quality	189	
Quality defects	160	
Major Maintenance	120	
	0	
	0	
	0	
	0	
	0	
	0	

Supply Chain Critical Items List

The supply chain participants are the raw material suppliers, the transportation providers, the production site, the customers. The rating titles may need to change or there may be a need for additional rating criteria.

The ability to list a specific area and have the supply chain participants evaluate the performance in the various areas provides a collaborative way of addressing issues. The issue discussion is probably more important than the rating.

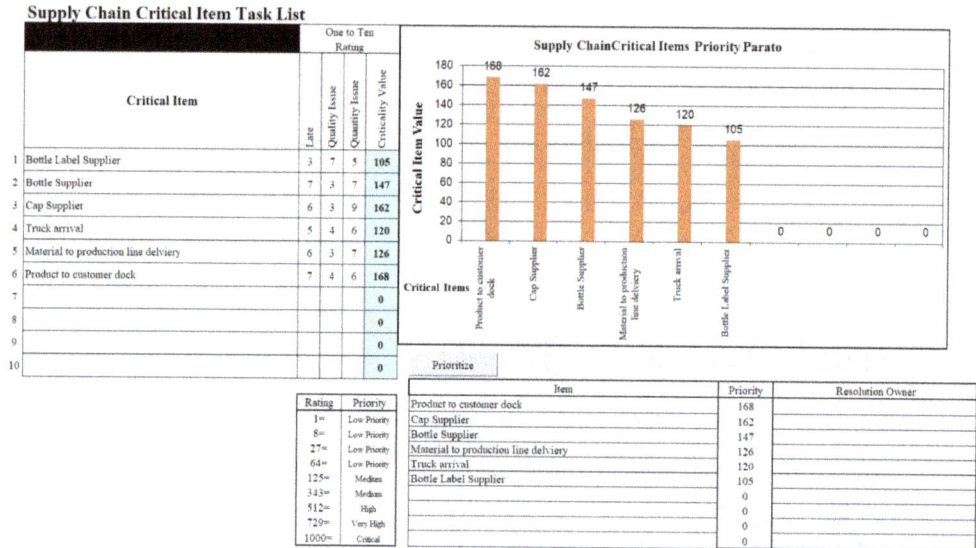

Supply Chain Critical Item Task List

Critical Item	Late	Quality Issue	Quantity Issue	Criticality Value
1 Bottle Label Supplier	3	7	5	105
2 Bottle Supplier	7	3	7	147
3 Cap Supplier	6	3	9	162
4 Truck arrival	5	4	6	120
5 Material to production line delivery	6	3	7	126
6 Product to customer dock	7	4	6	168
7				0
8				0
9				0
10				0

Supply Chain Critical Items Priority Parato

Critical Item Value: 168, 162, 147, 126, 120, 105, 0, 0, 0, 0

Critical Items: Product to customer dock, Cap Supplier, Bottle Supplier, Material to production line delivery, Truck arrival, Bottle Label Supplier

Rating	Priority
1=	Low Priority
8=	Low Priority
27=	Low Priority
64=	Low Priority
125=	Medium
343=	Medium
512=	High
729=	Very High
1000=	Critical

Prioritize

Item	Priority	Resolution Owner
Product to customer dock	168	
Cap Supplier	162	
Bottle Supplier	147	
Material to production line delivery	126	
Truck arrival	120	
Bottle Label Supplier	105	
	0	
	0	
	0	
	0	

Output Tracking

Output tracking is a powerful way to visualize the performance of an area that is producing a product or providing a service to the next in line.

Output Tracking											
Area: A+B+C+Applicator											
Team: Customization team						**Total Target Flow**	980				
Shift: Day shift						**Produced to Schedule**	980				
Area Leader: Maria Huyhn						**Produced Overall**	980				
Available Tim 7 hours						**Simple OEE**	100.0%			Center-Line	
Cycle Time 30 sec/person											
Unit/ Min 2 boxes											

Time hr.	Product	Target Flow	Actual Flow	Difference	Leader	Issue and Root Cause	Root Cause Responsibility	by Date	Stability Tracking	
									−	+
1	Facial Treatment Package	140	140	0						
2	Facial Treatment Package	140	135	(5)		Package scratch. Quality Issue				
3	Facial Treatment Package	140	140	0						
4	Facial Treatment Package	0	0	0						
5	Facial Treatment Package	140	140	0						
6	Facial Treatment Package	140	140	0						
7	Facial Treatment Package	140	140	0						
8	Facial Treatment Package	140	145	5						
9										
10										
11										
12										

Shift Summary
Five quality defect. Made up at last hour 8.
Shift ended with complete scheduled production

Output tracking used at every organizational boundary creates an organizational culture that basis its interactions on specific issues clarified by agreed to measures.

Maintenance Tasks List

1. Production Line Support

The maintenance organization team members participate in daily support meeting. They participate to understand the line and department need so they provide the needed support.

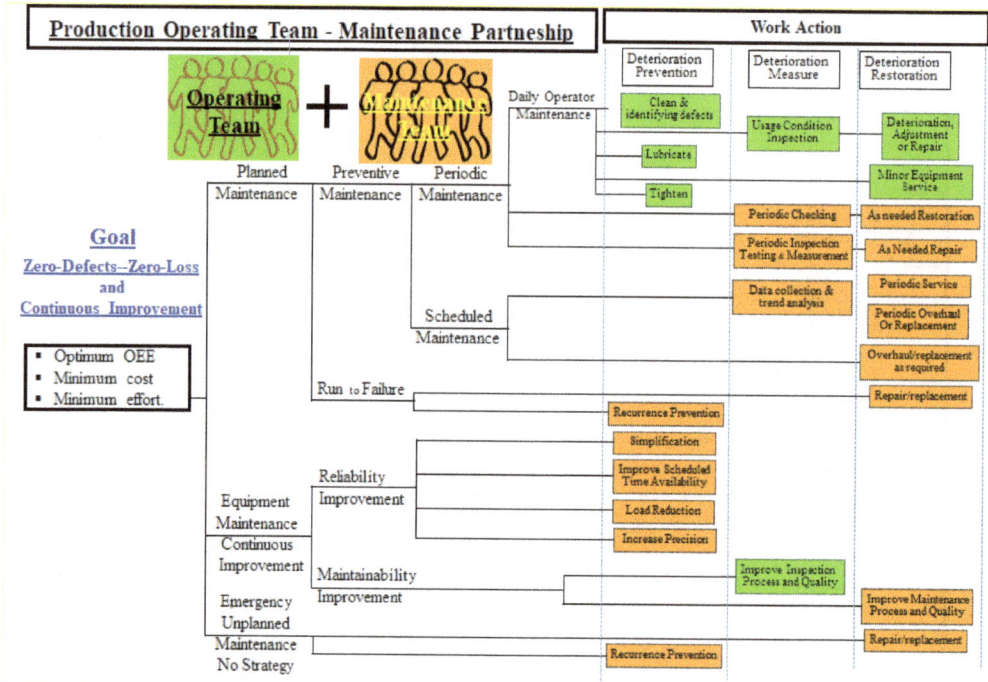

The operating teams establish a close working relationship and clearly identify the work actions each will do. Continuous equipment maintenance improvement to reduce lubrication and inspection time will be discussed and improvements planned.

Visual control standards are a key part of new improvements.

2. Break Down Counter Measures

The Maintenance department tracks breakdowns and categorizes them into Major, Moderate, and Minor breakdowns. This data is used as input to loss analysis and setting baseline data for goals to be set from.

Emphasis is placed on breakdown recurrence prevention. Failures are analyzed to root cause and operating standards, maintenance standards, and equipment improvements are put in place to reduce recurrence. Training is created to train technicians in the new standard procedures and equipment improvements are documented for future initiative work.

Standards and improvements are reapplied to similar equipment to prevent future failures.

As breakdowns occur, inspections are made of similar equipment to find developing failure situations so that potential breakdowns are immediately addressed.

Aligning rotating equipment, calibrating instruments, and maintaining control loop health are key activities.

Key components are inspected as a result of prevention activities.

As equipment components are maintained at basic and use conditions, poorly designed parts and their application will become apparent. Activities are deployed to make equipment modifications that will improve the life and extend of those components. Phenomena Mechanism Analysis are used to analyze failures due to design weaknesses and correct them

Establishing Maintenance Standards and Procedures

The maintenance standards that should be documented are:

- The standardized work processes
    - Equipment ranking criteria and process
    - Breakdown definition and classification
    - Process failure definition
    - Minor stop definition
    - Equipment logs
- A site wide numbering system for parts/materials, equipment & locations.
- The planning, scheduling, and budgeting maintenance process.
- Countermeasures to Breakdown Elimination daily management system
- Equipment Inspection visual controls
- Lubrication standards

- Planning and scheduling daily management system
- Alignment procedures for rotating equipment
- Instrument calibration and control loop health maintenance procedures
- Rebuild procedures
- Common standardized procedures are instrument calibration methods,

3. Pre-Breakdown Maintenance

The operating team's daily maintenance and operation activities greatly reduce unplanned maintenance and premature component failure. The maintenance resources focus on quality execution of corrective and periodic maintenance. The elimination of unplanned maintenance frees up time for resources to concentrate on developing cost-effective periodic maintenance systems supported by a computerized maintenance information system.

4. Facilities and tools management

- Special tools in place
- Adequate facilities for maintenance activities
- Availability of maintenance tools
- Tools and facilities are maintained at 5S standards.
- Tools and facilities are improved with a focus of reducing Mean time to repair (MTTR).

5. Equipment Priority Designation

- Identify which equipment on which to focus to improve in the areas of safety, quality, production, cost, delivery and materials.
- The equipment is ranked yearly
- Plans are modified to address any changes in business needs.
- Breakdown elimination activities and operator daily maintain activities are focused on the priority equipment
- Tracking systems compare the rate of improvement between priority equipment and non-priority equipment.
  Equipment priority designation is used to help establish which equipment and components will receive improvement focus.

6. <u>Maintenance Planning and Scheduling</u>

<u>Line Maintenance Schedule</u>

This example is presented as a way to show a two-year schedule of the required scheduled maintenance to maintain the line.

		Maintence Plan 2020												Maintence Plan 2021											
Operation 1	Equipment	January	February	March	April	May	June	July	August	September	October	November	December	January	February	March	April	May	June	July	August	September	October	November	December
Department 1	Line 1		Major OH														Major OH								
	Line 2			Major OH													Major OH								
	Line 3				Major OH												Major OH								
Department 2	Line 1					Major OH												Major OH							
	Line 2						Major OH												Major OH						
	Line 3							Major OH												Major OH					
Department 3	Line 1								Major OH												Major OH				
	Line 2									Major OH												Major OH			
	Line 3												Major OH											Major OH	
Operation 2	Equipment	January	February	March	April	May	June	July	August	September	October	November	December	January	February	March	April	May	June	July	August	September	October	November	December
Department 1	Line 1	Major OH												Major OH											
	Line 2		Major OH												Major OH										
	Line 3			Major OH												Major OH									
Department 2	Line 1				Major OH												Major OH								
	Line 2					Major OH												Major OH							
	Line 3						Major OH												Major OH						
Department 3	Line 1							Major OH													Major OH				
	Line 2								Major OH													Major OH			
	Line 3									Major OH													Major OH		

Major OH= Production Line Down for a minimum of one week
H = Production Line Down for 24 Hrs
s = Support as needed

This is used by the line team as they operate the production line and each level of the organization. The production planner integrates this information into the production schedule for each production line.

- Priorities for maintenance are set for next maintenance period.
- Related improvement projects are put into the plan.

7. <u>Lubrication Management system</u>

- Easy to maintain, optimized, simplified
- Transfusion clean lubrication to the required components.
- Conduct a lubrication survey
- Lubricant consolidation to the needed lubricants.
- Color standards and storage standards are created and are visually prominent.
- Production line operating teams are trained on the purpose, inspection methods, and application methods of lubrication. Maintenance resources provide tentative lubrication standards for the line operating teams and provide coaching and training as needed.

8. **Maintenance Information System**

Provide accurate and timely master data that enables maintenance planners to be more efficient and effective

The maintenance information system answers the following questions,

- What is your organization for master data creation?
- What are your change management processes?
- What % of parts is at the correct location for use by maintenance planners?
- How many days does it take to register a part?
- What are your focus areas?

9. **Maintenance and Equipment Cost system**

The maintenance and equipment cost system addresses the following,

- How much is spent on maintenance.
- The size of the maintenance budget.
- Identifies any repetitive maintenance tasks that:
a. Are particularly expensive.
b. Vary widely in cost each time they are executed.
- Identifies any equipment that costs more to maintain than similar equipment somewhere else.
- If maintenance spending is predictable.
- If you are on track to support the business need.
- Budget management is a focus in order to automate the analysis and reporting functions needed to control costs. The information used is:
  - Budget summaries,
  - work and materials usage schedules,
  - job priority lists,
  - equipment life forecasts, and
  - tracking maintenance costs by maintenance methods
    - Emergency,
    - Breakdown,
    - Time based maintenance,
    - Condition based maintenance.

10. <u>Parts and Supplies Control System</u>

- Parts and supply storage areas are straightened up
- Unnecessary parts are discarded.
- 5S methodology is used to organize the area and assign responsibilities for maintenance of the areas.
- A focus exists for quick retrieval of parts or supplies.
- Spare parts are categorized,
- storage methods are decided,
- Reorder methods are standardized.
- The focus is to reduce the spare part inventory to only the level that is needed to attain equipment availability goals and then reducing time required to find parts through visual control. A key success criterion is to find a needed part in less than 30 seconds.

Stress Free™ *Maintenance Solutions* has more details on all of these systems.

Individual Skill Spider Diagram

The goals for the maintenance area are,

- To achieve a 100%, line operation request response.
- To plan and schedule 90% of maintenance tasks thirty days in advance.
- To have 95% of maintenance tasks
- To have a maintenance strategy for all equipment
- To have a visual annual maintenance schedule.

The maintenance department has multiple responsibilities. There are eleven mentioned in this book. This list is not all inclusive.

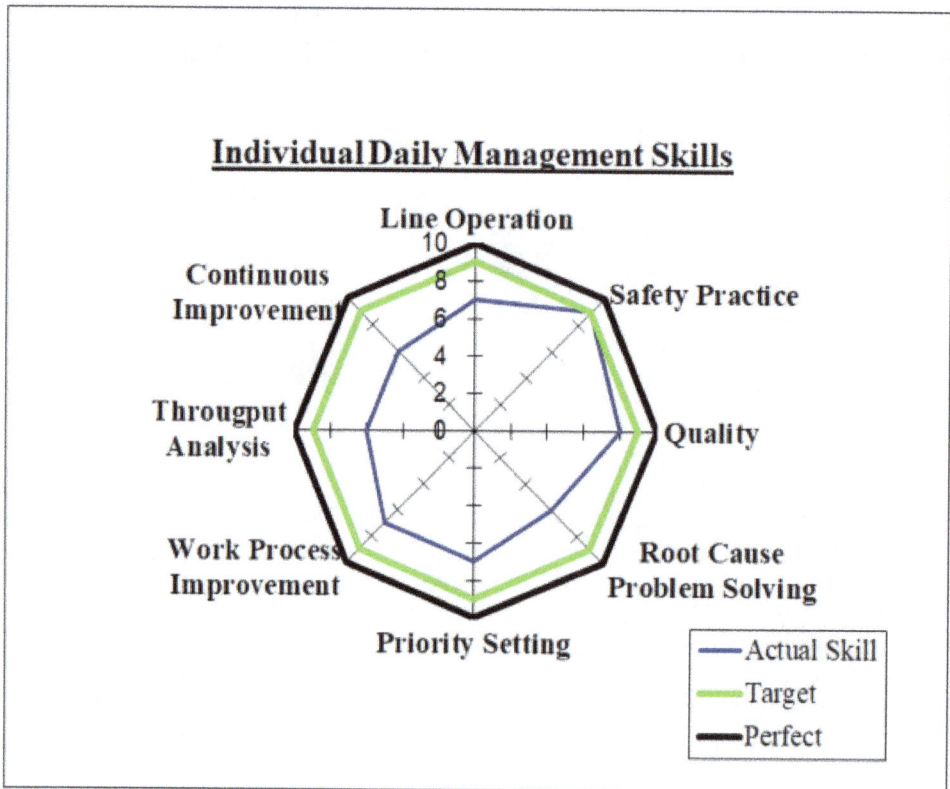

Individual Daily Management Skills

Line Team Skill Spider Diagram

A composited skill spider diagram applied to an operating team identifies the team's skill areas that need to be bolstered.

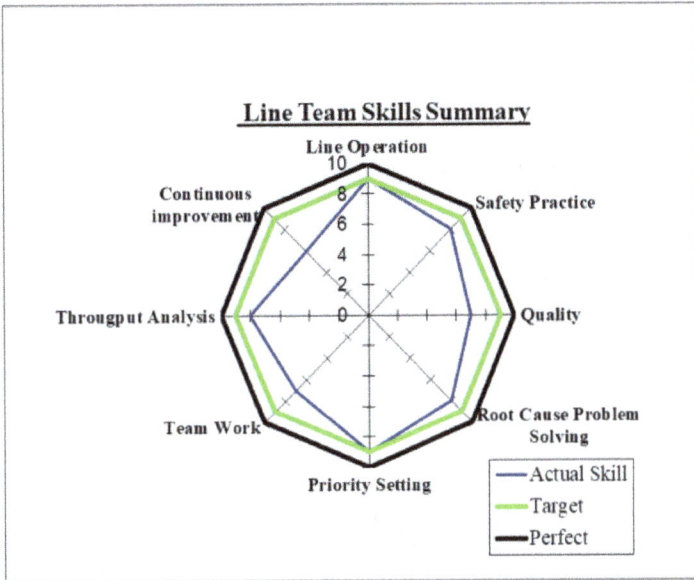

Line Team Skills Summary

Line Operation · Safety Practice · Quality · Root Cause Problem Solving · Priority Setting · Team Work · Throughput Analysis · Continuous improvement

— Actual Skill
— Target
— Perfect

Department Skills Summary

The combined skill diagrams of all the lines provide a visual indication of where the department should focus on increasing the department's skills.

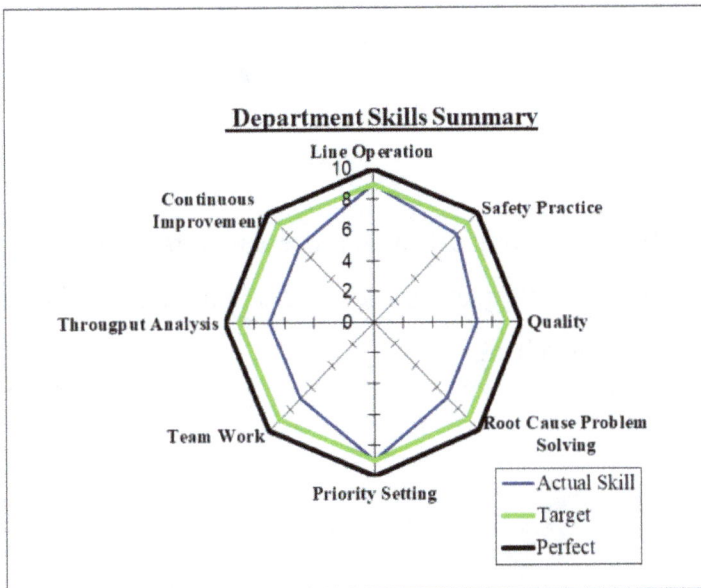

Department Skills Summary

Line Operation · Safety Practice · Quality · Root Cause Problem Solving · Priority Setting · Team Work · Throughput Analysis · Continuous Improvement

— Actual Skill
— Target
— Perfect

How to use for the total organization

The combining of skills up through the organization may be useful to almost any level. The choice of how high in the organization to go is dependent on the commonality of required skills. For instance, the required skills for support organizations may be significantly different than for the production area.

Organization Zero Loss Assessment

The assessment of losses across the organization can be quickly assessed by using a zero-loss assessment. This assessment when done with the key leaders of the organization requires no data other than what is in the head of each leader.

Organization Zero Loss Assessment

It is an eighty-five per cent effective process that may take as little as an hour to answer seventy five questions focused on thirteen parts of the organization and supply chain,

Ron Mueller P.E.

- Integrated Work Systems (IWS) materials author
- Coach to dozens of Manufacturing Directors across the world.
- Certified TPM Coach.
- Tested and proven to enable true breakthrough improvement of Supply Chains.

A proven leader of smart systems implementation across supply, manufacturing and distribution to drive out cost, inefficiencies and to establish synchronized Supply Chains. He utilized the best thinking of Japan's TPM leaders and crafted the necessary related pillars and systems that work in Consumer Products Manufacturing. The results delivered include reduction of Raw and Finished Product Inventories by 40%. Delivered over $100 million is loss reduction through focused systems Workshops across dozens of sites. Developed P&G IWS program materials for external sale. Winner of P&G's Diamond Award for Contribution to Product Supply.

Core Competencies include:

✓ Coaching Manufacturing Leadership,
✓ Implementation of Integrated Work Systems,
✓ Statistical Replenishment design and implementation,
✓ Supply Chain Synchronization, author of 3 books in the Stress Free™ series that aid Business and Supply Chain leaders to develop and improve their organization's performance.

Gordon Miller P.E.

- **Manufacturing Performance Program**
- **Development and Delivery Expert.**
- **Application of Intelligent Manufacturing technology against biggest business challenges with proven business results.**

A record as a collaborative and leading-edge thinker, developing programs to deliver cost, productivity and growth enabling manufacturing technology systems deployed via smart standards and empowered teams. As an early developer of PR/OEE measures and improvement programs, has experience with unlocking organization capability for improvement with smart strategies. Led program that developed initial P&G Manufacturing Execution System, leveraged globally across multiple GBUs. Influenced Beauty and Household Care manufacturing systems changes that enabled and leveraged global standardization for rapid footprint growth. Experience that enabled 50% reduction in OEE losses. Experience as a leader of corporate STEM talent strategy can assess and devise approaches to ensure Talent needs for the challenging future are met.

Core Competencies include:

- ✓ **Global Productivity Program Design and Management,**
- ✓ **Advanced Manufacturing Technology Innovation and Strategy Development,**
- ✓ **Development of Highly Effective Global Teams,**
- ✓ **Vendor development and management, Organization Capability Development,**
- ✓ **Talent Strategy**

Other books by Ron Mueller

Stress Free™ Supply Chain Solutions

Stress Free™ Manufacturing Solutions

Stress Free™ Work Process Solutions

Stress Free™ Changeover Solutions

Stress Free™ Daily Management Solutions

AROUND THE WORLD PUBLISHING, LLC,